T0289628

Finite Elements in Civil Engineering Applications

Finite Elements in Civil Engineering Applications

Justin Beil

WILLFORD PRESS
www.willfordpress.com

Published by Willford Press,
118-35 Queens Blvd., Suite 400,
Forest Hills, NY 11375, USA

ISBN: 978-1-64728-499-2

Cataloging-in-Publication Data

Finite elements in civil engineering applications / Justin Beil.
   p. cm.
Includes bibliographical references and index.
ISBN 978-1-64728-499-2
1. Finite element method. 2. Civil engineering. 3. Mechanics, Applied.
4. Structural engineering. 5. Engineering. I. Beil, Justin.
TA347.F5 F56 2023
620.001 515 35--dc23

For information on all Willford Press publications
visit our website at www.willfordpress.com

Contents

Permissions

Index

Preface

The purpose of the book is to provide a glimpse into the dynamics and to present opinions and studies of some of the scientists engaged in the development of new ideas in the field from very different standpoints. This book will prove useful to students and researchers owing to its high content quality.

Finite element analysis (FEA) is a tool used for numerical approximation of complex physical structures in the field of structural engineering. It is used for simulating physical phenomena in order to reduce dependency on the physical prototypes. This method allows optimization of the components as a part of the design process of the project. The simulations used in FEA are carried out by creating a mesh of a finite number of smaller elements. Thereafter, these finite elements integrate to form the shape of the structure that is being assessed. Each of these small elements is subjected to calculations, which are in the form of mathematical equations that predict the behavior of each element individually. A combination of such individual calculations produces the final result of the overall structure. FEA can be applied to areas such as structural analysis, heat transfer, mass transport and electromagnetic potential. This book is compiled in such a manner, that it will provide an in-depth knowledge about finite elements in civil engineering applications. Scholars and engineers in the field of civil engineering will be assisted by it.

At the end, I would like to appreciate all the efforts made by the authors in completing their chapters professionally. I express my deepest gratitude to all of them for contributing to this book by sharing their valuable works. A special thanks to my family and friends for their constant support in this journey.

Justin Beil

Introduction

1.1   Finite Element Method: An Overview

The finite element method is a numerical method for solving problems of engineering and mathematical physics. Typical problem areas of interest in engineering and mathematical physics that are solvable by use of the finite element method include structural analysis, heat transfer, fluid flow, mass transport and electromagnetic potential.

The problems involving complicated geometries, loadings and material properties, it is generally not possible to obtain analytical mathematical solutions. Analytical solutions are those given by a mathematical expression that yields the values of the desired unknown quantities at any location in a body and are thus valid for an infinite number of locations in the body.

These analytical solutions generally require the solution of ordinary or partial differential equations, which, because of the complicated geometries, loadings and material properties, are not usually obtainable. Hence we need to rely on numerical methods, such as the finite element method, for acceptable solutions.

The finite element formulation of the problem results in a system of simultaneous algebraic equations for solution, rather than requiring the solution of differential equations.

These numerical methods yield approximate values of the unknowns at discrete numbers of points in the continue. Hence this process of modeling a body by dividing it into an equivalent system of smaller bodies or units (finite elements) interconnected at points common to two or more elements (nodal points or nodes) and boundary lines and/or surfaces is called Discretization.

In the finite element method, instead of solving the problem for the entire body in one operation, we formulate the equations for each finite element and combine them to obtain the solution of the whole body.

Briefly, the solution for structural problems typically refers to determining the displacements at each node and the stresses within both elements making up the structure that is subjected to applied loads.

In nonstructural problems, the nodal unknowns may, for instance, be temperatures or fluid pressures due to thermal or fluid fluxes.

The method can easily handle factors such as non-linearity arbitrary loading conditions and time dependence. Any type of boundary condition can also be accommodated which is an important factor.

The Finite Element Method (FEM) is a numerical technique to find approximate solutions of partial differential equations. In a structural simulation FEM helps in producing stiffness and strength visualizations.

It also helps to minimize material weight and cost of the structures. FEM allows for detailed visualization and indicates the distribution of stresses and strains inside the body of a structure.

Most of the FE software are powerful yet complex tool meant for professional engineers with the training and education necessary to properly interpret the results. Many of modern FEM packages include specific components such as fluid, electromagnetic, thermal and structural working environments.

FEM allows entire designs to be constructed, refined and optimized before the design is manufactured. This powerful design tool has significantly improved both the standard of engineering designs and the methodology of the design process in many industrial applications.

The use of FEM has significantly decreased the time to take products from concept to the production line. One must take the advantage of the advent of faster generation of personal computers for the analysis and design of engineering product with precision level of accuracy.

1.2   Solution of Discrete Problems

Discretization is the process of subdividing the given body into a number of elements which results in a system of equivalent finite elements.

Discretization includes both node and element numbering, in this model every element connects two nodes so to distinguish between node numbering and element numbering elements numbers are encircled as shown in figure.

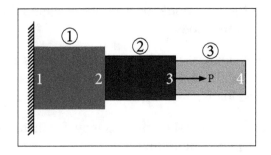

Above system can also be represented as a line segment as shown below:

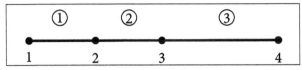

Discretization of element.

Here in one dimension every node is allowed to move only in one direction hence, each node as one degree of freedom. In the present case the model has four nodes it means four dots. Let Q_1, Q_2, Q_3 and Q_4 be the nodal displacements at node 1 to node 4 respectively similarly F_1, F_2, F_3, F_4 be the nodal force vector from node 1 to node 4 as shown.

When these parameters are represented for a entire structure use capitals which is called global numbering and for representing individual elements use small letters that is called local numbering as shown.

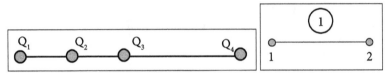

Global numbering local numbering.

This local and global numbering correspondence is established using element connectivity element as show in the figure below,

Elements	Nodes	
e	1	2
1	1	2
2	2	3
3	3	4

Element connectivity table.

Now let's consider a single element in a natural coordinate system that varies in ξ and η. x1 be the x coordinate of node 1 and x2 be the x coordinate of node 2 as shown below:

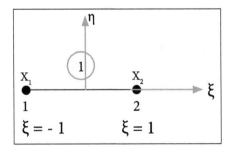

Let us assume a polynomial:

$$X = a_0 + a_1 \xi$$

Now,

At $X = X_1 \xi = -1$

At $X = X_2 \xi = 1$

After applying these conditions and solving for constants we have:

$$X_1 = a_0 - a_1$$

$$X_2 = a_0 + a_1$$

Substituting these constants in above equation we get:

$$a_0 = X_1 + X_2 / 2 a_1 = X_2 - X_1 / 2$$

Substituting these constants in above equation we get:

$$X = a_0 + a_1 \xi$$

$$X = \frac{X_1 + X_2}{2} + \frac{X_2 - X_1}{2} \xi$$

$$X = \frac{1 - \xi}{2} X_1 + \frac{1 + \xi}{2} X_2$$

$$X = N_1 X_1 + N_2 X_2$$

$$N_1 = \frac{1 - \xi}{2}, \qquad N_2 = \frac{1 + \xi}{2}$$

Where,

N_1 and N_2 are called shape functions also called as interpolation functions.

These shape functions can also be derived using nodal displacements say q1 and q2 which are nodal displacements at node 1 and node 2 respectively, now assuming the displacement function and following the same procedure as that of the nodal coordinate we get:

$$N_1 = \frac{1 - \xi}{2}, \qquad N_2 = \frac{1 + \xi}{2}$$

After applying these conditions and solving for constants we have:

$$X_1 = a_0 - a_1$$

$$X_2 = a_0 + a_1$$

$$U = \alpha_0 + \alpha_1 \xi$$

$$U = \frac{1-\xi}{2} q_1 + \frac{1+\xi}{2} q_2$$

$$U = N_1 q_1 + N_2 q_2$$

$$= [N_1 \ N_2] \begin{bmatrix} q_1 \\ q_2 \end{bmatrix}$$

$$U = Nq$$

Where,

N - Shape function matrix,

q - Displacement matrix.

Once the displacement is known its derivative gives strain and corresponding stress can be determined as follows:

$$U = Nq$$

$$\varepsilon = \frac{du}{dx} = \frac{du}{d\xi} \frac{d\xi}{dx}$$

$$\varepsilon = \frac{q_2 - q_1}{2} \frac{2}{x_2 - x_1}$$

$$\varepsilon = \frac{q_2 - q_1}{L}$$

where,

$$L = x_2 - x_1$$

$$\varepsilon = \frac{1}{L} [-1 \ 1] \begin{bmatrix} q_1 \\ q_2 \end{bmatrix}$$

$$\varepsilon = Bq$$

Where,

$$B = \frac{1}{L}[-1 \ 1], \text{ Element strain displacement matrix, } \sigma = E \, \varepsilon = B \, q \, E.$$

From the potential approach we have the expression of π as:

$$\pi = \frac{1}{2}\int_V \sigma^T \varepsilon \, dv - \int_V u^T f_b \, dv - \int_S u^T T \, ds - \sum_{i=1}^{n} u_i p_i$$

Since body is divided:

$$\pi_e = \int_e u_e - w_e dv$$

$$\pi = \frac{1}{2}\int B^T q^T E B q \, dv - \sum_{i=1}^{n} u_i p_i$$

Now total potential energy:

$$\pi = \sum \pi_e = \frac{1}{2}Q^T \left(\int B^T EBAL\right)Q - \sum Qi^T Fi$$

$$= 1/2 \, Q^T K Q - Q^T F$$

To differentiate the potential energy:

$$\frac{d\pi}{dQ^T} = 0 = KQ - F$$

1.2.1 Steady State and Time Dependent Continuous Problems

In this theory underlying the implementation of finite element methods in numerical analysis, particularly with reference to solutions of the partial differential equation:

$$-\nabla^2 u = f \text{ in } \Omega \qquad\qquad ...(1)$$

$$u = g_D \text{ on } \partial\Omega_D \text{ and } \frac{\partial u}{\partial n} = g_N \text{ on } \partial\Omega_N \qquad\qquad ...(2)$$

Where, $\Omega \subset R_2$, $\partial\Omega_D \cup \partial\Omega_N = \partial\Omega$ and the two boundary regions are distinct with $\partial\Omega_D$ non-empty. Specifically, we will consider the key results in the development of the theory of finite elements, including weak solutions and the structure of the individual elements, as well as the numerical implementation of the method.

We will subsequently implement a finite element numerical solver in order to solve this equation in a region of space, with f, gD and gN prescribed.

We will then extend these concepts in order to consider a time-varying problem using Euler time-steps and finally implement a finite element solution to a simple time-varying analogue of the original spatial partial differential equation, given as:

$$\frac{\partial u}{\partial t} - \nabla^2 u = f \text{ in } \Omega \qquad \qquad \text{...(3)}$$

$$u = u_0 \text{ in } \Omega \text{ at } t = 0 \qquad \qquad \text{...(4)}$$

With the same boundary conditions as before. We will consider the problem with both steady and time-varying boundary conditions.

Time Dependent Continuous

In a general time-dependent or propagation problem, three spatial and one time parameters will be involved. Usually, we first use the finite element method to formulate the solution in the physical space. Next, we use a different method, such as finite differences, to find the solution over a period of time. Thus, this procedure involves the idealization of the field variable ϕ (x, y, z, t) in any element e (in three-dimensional space) as:

$$\phi \, (x, y, z, t) = \left[N_1 (x, y, z,) \right] \vec{\Phi}^{(e)} (t) \qquad \qquad \text{...(1)}$$

Where, [N1] is the matrix of interpolation or shape functions in space, and $\vec{\Phi}^{(e)}$ is the vector of time-dependent nodal variables. By using equation.(1) and the specified initial conditions, we use a finite difference scheme such as,

$$\vec{\Phi}^{(e)} (t) = \left[N_2 (t, \Delta t) \right] \vec{\Phi}^{(e)} (t - \Delta t) \qquad \qquad \text{...(2)}$$

Where, [N₁] indicates the matrix of interpolation functions in the time domain. Instead of solving the problem using equation. (1) And (2), finite elements can be constructed in four dimensions (x, y, z, and t) and the field variable can be expressed as follows:

$$\phi \, (x, y, z, t) = \left[N_1 (x, y, z, t) \right] \vec{\Phi}^{(e)} \qquad \qquad \text{...(3)}$$

Where, [N] represents the matrix of shape functions in x, y, z, and t, and $\vec{\Phi}^{(e)}$ is the vector of nodal values of element e. In this case, the time-dependent problem can be directly solved without using any special techniques.

Finite Element Methods on a Spatial Problem

Weak Formulation

Many equations of interest in applied mathematics have 'classical' solutions. We consider the sets of functions:

$$C^0(\Omega)=\{f:\Omega \to R \,|\, f \text{ is continuous}\},$$

$$C^2(\Omega)=\{f:\Omega \to R \,|\, fx_i x_j \in C^0(\Omega) \forall \text{ i, j}\}.$$

If Ω is sufficiently regular, the solution u to an second-order partial differential equation, such as that given in (1, 2) satisfying u $\in C^2(\Omega)$ is known as a classical or strong solution. That is, if there exists a solution which has continuous second partial derivatives and is therefore, sufficiently smooth, it is a strong solution to the problem.

There are numerous numerical techniques that may be applied in order to find classical solutions to partial differential equations. Unfortunately, many problems of interest do not admit classical solutions.

If this is the case, we must instead search for a variational or weak formulation of the original problem. This involves reformulating the original problem such that the smoothness requirement is relaxed. The resultant expression is known as a weak formulation of the problem.

In order to obtain a weak formulation, we consider a test function ω that lies in the sobolev space:

$$H^1_{E_0}(\Omega)=\{v \in H^1(\Omega) \,|\, v=0 \text{ on} \partial\Omega_D\} \qquad \text{...(4)}$$

Where,

$$H^1(\Omega)=\{v:\Omega \to \mathbb{R} \,|\, u, u_x \, u_y \in L_2(\Omega)\}$$

$$L_2(\Omega)=\left\{v:\Omega \to \mathbb{R} \,\Big|\, \int_\Omega v^2 < \infty\right\}$$

That is, we require that the function ω and associated first-derivatives to be square integrable. Now, if we integrate the product of problem given in (1) and the test function ω, we obtain:

$$-\int_\Omega \nabla^2 u \cdot \omega = \int_\Omega f \cdot \omega$$

Integration by parts and application of the divergence theorem gives:

$$-\int_\Omega \nabla u \cdot \nabla\omega = \int_\Omega \omega f + \int_{\partial\Omega} \omega \frac{\partial u}{\partial n}.$$

Substitution of the boundary conditions given in (2), noting that $v \in H^1_{E_0}(\Omega)$, gives the formulation.

$$\int_\Omega \nabla u \cdot \nabla \omega = \int_\Omega \omega f + \int_{\partial \Omega_N} \omega g_N \qquad \ldots (5)$$

This is the weak formulation of the partial differential equation given in (1). In order for a solution u to be considered a weak solution of (1, 2), we require that:

$$u \in H^1_E(\Omega) = \left\{ v \in H^1(\Omega) \,|\, v = g_D \text{ on } \partial \Omega_D \right\}$$

and that u satisfies the weak formulation of (1), given in (5). While for strong solutions we required the second-derivatives to be continuous, we simply require for weak solutions that the first-derivatives be square-integrable.

It may be demonstrated that all strong solutions to the problem also satisfy the conditions to be considered weak solutions. By considering the weak formulation, we are effectively permitting a broader range of solutions to the original problem by extending the solution space.

Galerkin Method

The Galerkin approximation method allows us to convert a continuous problem, such as the weak formulation for the partial differential equation into a discrete the problem that may be solved numerically.

This method involves using a finite dimensional subspace So of $H^1_{E_0}$ and therefore obtaining a solution in a finite dimensional subset of H^1_E.

The first stage is to tesselate the domain Ω into non-overlapping regions. Any polygonal domain may be tessellated using triangles and as such, triangular tessellations are frequently chosen when implementing the Galerkin method.

It is, however, possible to divide the domain up in a variety of fashions, such as quadrilateral tessellations. We will assume that the domain is tessellated with non-overlapping triangles or elements Δ_k for k = 1,..., K, such that the vertices of neighboring elements meet at points called nodes. An example of this tessellation is included in the figure (a).

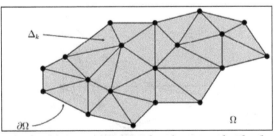

(a) Triangular tesselation of a domain Ω using triangular elements Δk. The domain is shaded blue, with nodes denoted by bold circles and the boundary of the domain, $\partial \Omega$, denoted by bold lines.

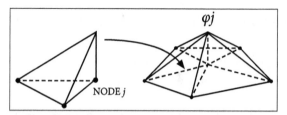

(b) The right-hand side shows a schematic of a typical φj, which is nonzero at node j and zero at all other nodes, with φj defined in a piecewise fashion such that it is linear in each triangle. The left-hand side shows φj on a single triangle. Each triangle will be nonzero for three different values of j, corresponding to each of the three nodes on the edge of the triangle.

The subspace S_o is typically defined as the span of a set of basic functions, such that:

$$S_o = \text{span}\left\{\phi_1, \phi_2, ..., \phi_n\right\},$$

Where, φj for j = 1,..., n is a set of linearly independent basis functions and n is the number of nodes in the tesselation of the domain that are not found on $\partial\Omega_D$. The remaining nodes found on $\partial\Omega D$ are denoted as node $j = n+1, n+2,...,n+n\partial$.

A common method of constructing So is to select φj such that φj is continuous, equal to one at node j and equal to zero on all elements not touching node j. On the elements that do touch node j, φj is linear in the form ax + by + c such that the previous requirements are satisfied.

A schematic of the function φj shown in the figure (b). The set $\left\{\phi_1, \phi_2,...,\phi_n\right\}$ forms a basis for the space So = P1, which is the space of continuous piecewise linear functions on the triangulation. We also define $\phi_{n+1}, \phi_{n+2},...,\phi_{n+n\partial}$ in a similar fashion.

There are other methods of constructing the approximation space So such that it takes different forms. It is possible to use quadratic functions of the form $ax^2 + bxy + cy^2 + dx + ey + f$, rather than linear functions, on each element.

This requires nodes to be defined on the sides of each triangle, as well as at the vertices. In fact, it is possible to select polynomials of any order, although this typically involves selecting more nodes on the boundary, as well as internal nodes.

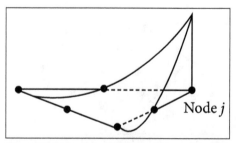

(c): A single triangular element containing part of a piecewise quadratic basis function φj. The element contains six nodes, which is required in order to generate the basis function.

To define cubic basis functions, we require that the triangles contain 10 nodes, with four on each side and one internal node. While there are some situations where selecting higher-order basis functions is desirable and we will consider only linear basis functions for the purposes of calculation.

For illustrative purposes, figure (c) contains an example of a triangular element containing part of a quadratic basis function. Additionally, the elements may not be triangular, but may instead take different shapes, such as quadrilateral elements. Again, there are some situations where such basic functions may be more useful for computational purposes.

The Galerkin method involves seeking:

$$u_h = \sum_{j=1}^{n} u_j \phi_j + \sum_{j=n+1}^{n+n\partial} u_j \phi_j \in H_E^1 \qquad \qquad ...(6)$$

Such that uh satisfies the weak form of the original problem, given in equation (5). This gives,

$$\int_\Omega \nabla u_h \cdot \nabla \omega = \int_\Omega \omega f + \int_{\partial\Omega_N} \omega g_N \qquad \qquad ...(7)$$

For all $\omega \in S_o$. As $S_o = \text{span}\{\phi_1, \phi_2, ..., \phi_n\}$, this is equivalent to finding uh such that,

$$\int_\Omega \nabla u_h \cdot \nabla \phi_i = \int_\Omega \phi_i f + \int_{\partial\Omega_N} \phi_i g_N$$

For i = 1, 2,..., n. Using equation (6), this may be expressed in the form of a linear system $Au = f$ such that,

$$A = \{a_{i,j}\}$$

where,

$$a_{i,j} = \int_\Omega \nabla\phi_j \cdot \nabla\phi_i \qquad \qquad ...(8)$$

$$u = (u_1, u_2, ..., u_n)^T \text{ and } f = (f_1, f_2, ..., f_n)^T,$$

where,

$$f_i = \int_\Omega \phi_i f + \int_{\partial\Omega_N} \phi_i g_N - \sum_{j=n+1}^{n+n\partial} u_j \int_\Omega \nabla\phi_j \cdot \nabla\phi_i$$

As such, the Galerkin method involves determining A and f and then solving the resultant linear system in order to obtain an approximation for u in So. We could apply this method to problems with different governing equations, however this entails obtaining a new weak formulation for the problem.

1.3   Applications of Finite Method through Illustrative Examples

The finite element method can be used to analyze both structural and nonstructural problems.

Typical Structural Areas Include:

- Stress analysis, including truss and frame analysis and stress concentration problems typically associated with holes, fillets or other changes in geometry in a body.

- Buckling.

- Vibration analysis.

Nonstructural Problems Include:

- Heat transfer.

- Fluid flow, including seepage through porous media.

- Distribution of electric or magnetic potential.

Finally, some biomechanical engineering problems (which may include stress analysis) typically include analyses of human spine, skull, hip joints, jaw/gum tooth implants, heart and eye.

Examples:

Control Tower for a Railroad

We now present some typical applications of the finite element method. These applications will illustrate the variety, size and complexity of problems that can be solved using the method and the typical Discretization process and kinds of elements used.

The figure (a) illustrates a control tower for a railroad. The tower is a three-dimensional frame comprising a series of beam-type elements. The 48 elements are labeled by the circled numbers, whereas the 28 nodes are indicated by the un-circled numbers.

Each node has three rotation and three displacement components associated with it. The rotations (θs) and displacements (ds) are called the degrees of freedom. Because of the loading conditions to which the tower structure is subjected, we have used a three-dimensional model.

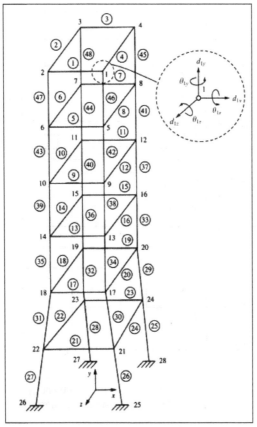

Figure (a) Discretized railroad control tower (28 nodes, 48 beam elements) with typical degrees of freedom shown at node 1.

The finite element method used for this frame enables the designer/analyst quickly to obtain displacements and stresses in the tower for typical load cases, as required by design codes.

Underground Box Culvert

The next illustration of the application of the finite element method to problem solving is the determination of displacements and stresses in an underground box culvert subjected to ground shock loading from a bomb explosion.

Figure (b) shows the discretized model, which included a total of 369 nodes, 40 one-dimensional bar or truss elements used to model the steel reinforcement in the box culvert and 333 plane strain two-dimensional triangular and rectangular elements used to model the surrounding soil and concrete box culvert.

With an assumption of symmetry, only half of the box culvert need be analyzed. This problem requires the solution of nearly 700 unknown nodal displacements.

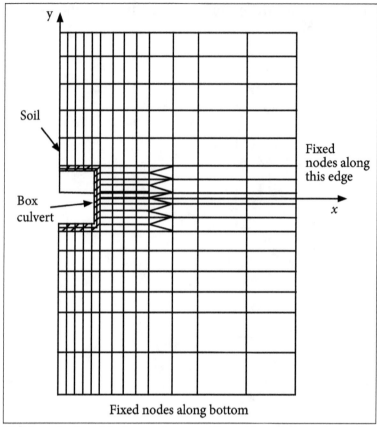

(b) Discretized model of an underground box culvert (369 nodes, 40 bar elements and 333 plane strain elements).

It illustrates that different kinds of elements (here bar and plane strain) can often be used in one finite element model.

Hydraulic Cylinder Rod End

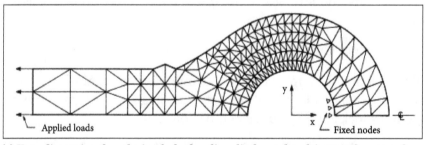

(c) Two-dimensional analysis of a hydraulic cylinder rod end (120 nodes, 297 plane strain triangular elements).

Another problem, that of the hydraulic cylinder rod end shown in the figure(c), was modeled by 120 nodes and 297 plane strain triangular elements. Symmetry was also applied to the whole rod end so that only half of the rod end had to be analyzed, as shown.

The purpose of this analysis was to locate areas of high stress concentration in the rod end.

Chimney Stack Section

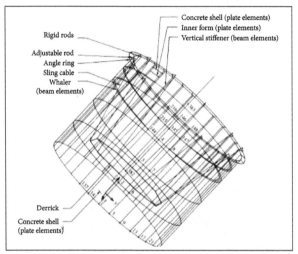

(d) Finite element model of a chimney stack section (end view rotated 45°)
(584 beam and 252 flat-plate elements).

Figure (d) shows a chimney stack section that is four form heights (or a total of 32 ft high). In this illustration, 584 beam elements were used to model the vertical and horizontal stiffeners making up the formwork and 252 flat-plate elements were used to model the inner wooden form and the concrete shell.

Because of the irregular loading pattern on the structure, a three-dimensional model was necessary. Displacements and stresses in the concrete were of prime concern in this problem.

Steel Die

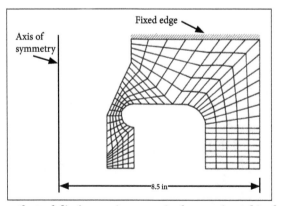

(e) Model of a high-strength steel die (240 axisymmetric elements) used in the plastic film industry.

The figure (e) shows the finite element discretized model of a proposed steel die used in a plastic film-making process. The irregular geometry and associated potential stress concentrations necessitated use of the finite element method to obtain a reasonable solution. Here 240 axisymmetric elements were used to model the three-dimensional die.

Three-Dimensional Solid Element

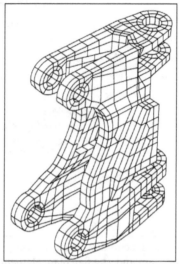

(f) Three-dimensional solid element model of a swing casting for a backhoe frame.

Figure (f) illustrates the use of a three-dimensional solid element to model a swing casting for a backhoe frame. The three-dimensional hexahedral elements are necessary to model the irregularly shaped three-dimensional casting. Two-dimensional models certainly would not yield accurate engineering solutions to this problem.

Two-Dimensional Heat-Transfer Model

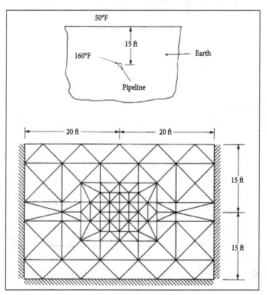

(g) Finite element model for a two-dimensional temperature distribution in the earth.

The figure (g) illustrates a two-dimensional heat-transfer model used to determine the temperature distribution in earth subjected to a heat source a buried pipeline transporting a hot gas.

Three-Dimensional Finite Element Model of A Pelvis Bone

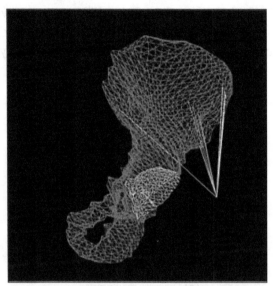

(h) Finite element model of a pelvis bone with an implant (over 5000 solid elements were used in the model).

The figure (h) shows a three-dimensional finite element model of a pelvis bone with an implant, used to study stresses in the bone and the cement layer between bone and implant.

Three-Dimensional Model of a 710G Bucket

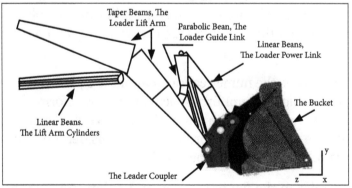

(i) Finite element model of a 710G bucket with 169,595 elements and 185,026 nodes used (including 78,566 thin shell linear quadrilateral elements for the bucket and coupler, 83,104 solid linear brick elements to model the bosses and 212 beam elements to model lift arms, lift arm cylinders and guide links).

Finally, Figure (i) shows a three-dimensional model of a 710G bucket, used to study stresses throughout the bucket. These illustrations suggest the kinds of problems that can be solved by the finite element method.

1.4   Finite Difference Method

- Analytical solutions of partial differential equations provide us with closed-form expressions which depict the variation of the dependent variables in the domain.

- The numerical solutions, based on finite differences, provide us with the values at discrete points in the domain which are known as grid points. Consider the figure (a), which shows a domain of calculation in the x–y plane.

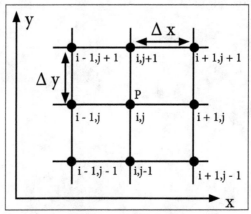

(a) Discrete grid points.

Let us assume that the spacing of the grid points in the x–direction is uniform and given by Δx. Likewise, the spacing of the points in the y–direction is also uniform and given by Δy. It is not necessary that Δx or Δy be uniform.

We could imagine unequal spacing in each direction, where different values of Δx between both successive pairs of grid points are used. The same could be presumed for Δy as well. However, often, problems are solved on a grid which involves uniform spacing in both directions, because this simplifies the programming and often results in higher accuracy.

In some class of problems, the numerical calculations are performed on a transformed computational plane which has uniform spacing in the transformed independent variables, but non-uniform spacing in the physical plane. We shall consider uniform spacing in both coordinate directions. According to our consideration, Δx and Δy are constants, but it is not mandatory that Δx be equal to Δy.

Let us once again refer to the figure (a). The grid points are identified by an index i which increases in the positive x-direction and an index j, which increases in the positive y-direction.

If (i, j) is the index of point P in the figure (a), then the point immediately to the right is designated as $(i+1, j)$ and the point immediately to the left is $(i-1, j)$. Similarly the point directly above is $(i, j+1)$ and the point directly below is $(i, j-1)$.

The basic philosophy of finite difference methods is to replace the derivatives of the governing equations with algebraic difference quotients. This will result in a system of algebraic equations which can be solved for the dependent variables at the discrete grid points in the flow field.

Let us now look at some of the common algebraic difference quotients in order to be acquainted with the methods related to discretization of the partial differential equations.

Elementary Finite Difference Quotients

Finite difference representations of derivatives are derived from Taylor series expansions. For example, if $u_{i,j}$ is the x–component of the velocity $u_{i+1,j}$ at point $(I + 1, j)$ can be expressed in terms of Taylor series expansion about point (i, j) as,

$$u_{i+1j} = u_{i,j} + \left(\frac{\partial u}{\partial x}\right)_{i,j} \Delta x + \left(\frac{\partial^2 u}{\partial x^2}\right)_{i,j} \frac{(\Delta x)^2}{2} + \left(\frac{\partial^3 u}{\partial x^3}\right)_{i,j} \frac{(\Delta x)^3}{6} + \dots \quad \dots(1)$$

Mathematically, equation (1) is an exact expression for $u_{i+1,j}$ if the series converges. In practice, Δx is small and any higher-order term of Δx is smaller than Δx. Hence, for any function $u(x)$, the equation (1) can be truncated after a finite number of terms. For example, if terms of magnitude $(\Delta x)3$ and higher order are neglected, equation (1) becomes,

$$u_{i+1j} \approx u_{i,j} + \left(\frac{\partial u}{\partial x}\right)_{i,j} \Delta x + \left(\frac{\partial^2 u}{\partial x^2}\right)_{i,j} \frac{(\Delta x)^2}{2} \quad \dots(2)$$

The equation (2) is second-order accurate, because terms of order $(\Delta x)3$ and higher have been neglected. If terms of order $(\Delta x)2$ and higher are neglected, equation (2) is reduced to,

$$u_{i+1j} \approx u_{i,j} + \left(\frac{\partial u}{\partial x}\right)_{i,j} \Delta x \quad \dots(3)$$

Equation (3) is first-order accurate. In equations (2) and (3) the neglected higher order terms represent the truncation error. Therefore, the truncation errors for equations (2) and (3) are:

$$\sum_{n=3}^{\infty} = \left(\frac{\partial^n u}{\partial x^n}\right)_{i,j} \frac{(\Delta x)^n}{n!}$$

And

$$\sum_{n=2}^{\infty} = \left(\frac{\partial^n u}{\partial x^n}\right)_{i,j} \frac{(\Delta x)^n}{n!}$$

It is now obvious that the truncation error can be reduced by retaining more terms in the Taylor series expansion of the corresponding derivative and reducing the magnitude of Δx. Let us once again return to equation (1) and solve for $(\partial u / \partial x)_{i,j}$ as:

$$\left(\frac{\partial u}{\partial x}\right)_{i,j} = \frac{u_{i+1,j}}{\Delta x} - \left(\frac{\partial^2 u}{\partial x^2}\right)_{i,j}\frac{\Delta x}{2} - \left(\frac{\partial^3 u}{\partial x^3}\right)_{i,j}\frac{(\Delta x)^2}{6} + \dots$$

Or

$$\left(\frac{\partial u}{\partial x}\right)_{i,j} = \frac{u_{i+1,j} - u_{i,j}}{\Delta x} + O(\Delta x) \qquad \dots(4)$$

In equation (4) the symbol $O(\Delta x)$ is a formal mathematical nomenclature which means "terms of order of Δx", expressing the order of magnitude of the truncation error.

The first-order-accurate difference representation for the derivative $(\partial u / \partial x)_{i,j}$ expressed by equation (4) can be identified as a first-order forward difference. We now consider a Taylor series expansion for $u_{i-1,j}$, about $u_{i,j}$.

$$u_{i-1,j} = u_{i,j} + \left(\frac{\partial u}{\partial x}\right)_{i,j}(-\Delta x) + \left(\frac{\partial^2 u}{\partial x^2}\right)_{i,j}\frac{(-\Delta x)^2}{2} + \left(\frac{\partial^3 u}{\partial x^3}\right)_{i,j}\frac{(-\Delta x)^3}{6} + \dots$$

Or

$$u_{i-1,j} = u_{i,j} - \left(\frac{\partial u}{\partial x}\right)_{i,j}(\Delta x) + \left(\frac{\partial^2 u}{\partial x^2}\right)_{i,j}\frac{(\Delta x)^2}{2} - \left(\frac{\partial^3 u}{\partial x^3}\right)_{i,j}\frac{(\Delta x)^3}{6} + \dots(5)$$

Solving for $(\partial u / \partial x)_{i,j}$, we obtain:

$$\left(\frac{\partial u}{\partial x}\right)_{i,j} = \frac{u_{i,j} - u_{i-1,j}}{\Delta x} + O(\Delta x) \qquad \dots(6)$$

Equation (6) is a first-order backward expression for the derivative $(\partial u / \partial x)$ at grid point (i, j).

Subtracting equation (5) from (1), we get:

$$u_{i+1,j} - u_{i-1,j} = 2\left(\frac{\partial u}{\partial x}\right)_{i,j}(\Delta x) + \left(\frac{\partial^3 u}{\partial x^3}\right)_{i,j}\frac{(\Delta x)^3}{3} + \dots \qquad \dots(7)$$

And solving for $(\partial u / \partial x)_{i,j}$ from equation (7) we obtain,

$$\left(\frac{\partial u}{\partial x}\right)_{i,j} = \frac{u_{i+1,j} - u_{i-1,j}}{2\Delta x} + O(\Delta x)^2 \qquad \dots(8)$$

Equation (8) is a second-order central difference for the derivative $(\partial u / \partial x)$ at grid point (i, j). In order to obtain a finite-difference for the second-order partial derivative $\left(\partial^2 u / \partial x^2\right)_{i,j}$, add equation (1) and (5). This produces,

$$u_{i+1,j} + u_{i-1,j} = 2u_{i,j} + \left(\frac{\partial^2 u}{\partial x^2}\right)_{i,j} (\Delta x)^2 + \left(\frac{\partial^4 u}{\partial x^4}\right)_{i,j} \frac{(\Delta x)^4}{12} + \ldots \qquad \ldots(9)$$

Solving equation (9) for $\left(\partial^2 u / \partial x^2\right)_{i,j}$, we obtain

$$\left(\frac{\partial^2 u}{\partial x^2}\right)_{i,j} = \frac{u_{i+1,j} - 2u_{i-1,j}}{(\Delta x)^2} + O(\Delta x)^2 \qquad \ldots(10)$$

Equation (10) is a second-order central difference form for the derivative $\left(\partial^2 u / \partial x^2\right)$ at grid point (i, j). Difference quotients for the y-derivatives are obtained in exactly the similar way. The results are analogous to the expressions for the x-derivatives.

$$\left(\frac{\partial}{\partial y}\right)_{i,j} = \frac{u_{i,j+1}\;u_{i,j}}{\Delta y} + O(\Delta y) \quad \text{[Forward difference]}$$

$$\left(\frac{\partial u}{\partial y}\right)_{i,j} = \frac{u_{i,j} - u_{i,j-1}}{\Delta y} + O(\Delta y) \quad \text{[Backward difference]}$$

$$\left(\frac{\partial u}{\partial y}\right)_{i,j} = \frac{u_{i,j+1} - u_{i,j-1}}{2\Delta y} + O(\Delta y)^2 \quad \text{[Central difference]}$$

$$\left(\frac{\partial^2 u}{\partial y^2}\right)_{i,j} = \frac{u_{i,j+1} - 2u_{i,j} + u_{i,j-1}}{(\Delta y)^2} + O(\Delta y)^2 \quad \text{[Central difference of second derivative]}$$

It is interesting to note that the central difference given by equation (10) can be interpreted as a forward difference of the first order derivatives, with backward differences in terms of dependent variables for the first-order derivatives. This is because,

$$\left(\frac{\partial^2 u}{\partial x^2}\right)_{i,j} = \left[\frac{\partial}{\partial x}\left(\frac{\partial u}{\partial x}\right)\right]_{i,j} = \frac{\left(\frac{\partial u}{\partial x}\right)_{i+1,j} - \left(\frac{\partial u}{\partial x}\right)_{i,j}}{\Delta x}$$

Or

$$\left(\frac{\partial^2 u}{\partial x^2}\right)_{i,j} = \left[\left(\frac{u_{i+1,j} - u_{i,j}}{\Delta x}\right) - \left(\frac{u_{i,j} - u_{i-1,j}}{\Delta x}\right)\right]\frac{1}{\Delta x}$$

Or

$$\left(\frac{\partial^2 u}{\partial x^2}\right)_{i,j} = \frac{u_{i+1,j} - 2u_{i,j} + u_{i-1,j}}{(\Delta x)^2}$$

The same approach can be made to generate a finite difference quotient for the mixed derivative $\left(\partial^2 u / \partial x \partial y\right)$ at grid point (i, j). For example:

$$\frac{\partial^2 u}{\partial x \partial y} = \frac{\partial}{\partial x}\left(\frac{\partial u}{\partial y}\right) \qquad \qquad ...(11)$$

In equation (11), if we write the x–derivative as a central difference of y-derivatives and further make use of central differences to find out the y–derivatives, we obtain

$$\frac{\partial^2 u}{\partial x \partial y} = \left[\left(\frac{u_{i+1,j+1} - u_{i+1,j+1}}{2(\Delta y)}\right) - \left(\frac{u_{i-1,j+1} - u_{i-1,j-1}}{2(\Delta y)}\right)\right]\frac{1}{2(\Delta x)}$$

$$\frac{\partial^2 u}{\partial x \partial y} = \frac{\partial}{\partial x}\left(\frac{\partial u}{\partial y}\right) = \frac{\left(\frac{\partial u}{\partial y}\right)_{i+1,j} - \left(\frac{\partial u}{\partial y}\right)_{i-1,j}}{2(\Delta x)}$$

$$\left(\frac{\partial^2 u}{\partial x \partial y}\right) = \frac{1}{4\,\Delta x\,\Delta y}\left(u_{i+1,j+1} + u_{i-1,j-1} - u_{i+1,j-1} - u_{i-1,j+1} + \right.$$

$$\left. O\left[(\Delta x)^2, (\Delta y)^2\right]\right) \qquad \qquad ..(12)$$

Combinations of such finite difference quotients for partial derivatives form finite difference expressions for the partial differential equations. For example, the Laplace equation $\nabla 2u = 0$ in two dimensions becomes:

$$\frac{u_{i+1,j} - 2u_{i,j} + u_{i-1,j}}{(\Delta x)^2} + \frac{u_{i,j+1} - 2u_{i,j} + u_{i,j-1}}{(\Delta y)^2} = 0$$

Or

$$u_{i+1,j} + u_{i-1,j} + \lambda^2\left(u_{i,j+1} + u_{i,j-1}\right) - 2\left(1+\lambda^2\right)u_{i,j} = 0 \qquad ...(13)13$$

Where, λ is the mesh aspect ratio $(\Delta x)/ (\Delta y)$. If we solve the Laplace equation on a domain given by the figure (a), the value of ui,j will be,

$$u_{i,j} = \frac{u_{i+1,j} + u_{i-1,j} + \lambda^2\left(u_{i,j+1} + u_{i,j-1}\right)}{2\left(1+\lambda^2\right)} \qquad \qquad ...(14)$$

It can be said that many other forms of difference approximations can be obtained for the derivatives which constitute the governing equations for fluid flow and heat transfer. The basic procedure, however, remains the same.

1.5 Representation of Differential Equations: Stability, Consistency and Convergence

Consider the following one dimensional unsteady state heat conduction equation. The dependent variable u (temperature) is a function of x and t (time) and α is a constant known as thermal diffusivity.

$$\frac{\partial u}{\partial t} = \alpha \frac{\partial^2 u}{\partial x^2} \qquad \qquad ...(1)$$

It is to be noted that equation (1) is classified as a parabolic partial differential equation.

If we substitute the time derivative in equation (1) with a forward difference and a spatial derivative with a central difference (usually called FTCS, Forward Time Central Space method of discretization), we obtain

$$\frac{u_i^{n+1} - u_i^n}{\Delta t} = \alpha \left[\frac{u_{i+1}^n - 2u_i^n + u_{i-1}^n}{\left(\Delta x^2\right)} \right] \qquad \qquad ...(2)$$

In equation (2), the index for time appears as a superscript:

Where, n denotes conditions at time t, (n+ 1) denotes conditions at time (t+Δt) and so on. The subscript denotes the grid point in the spatial dimension.

However, there must be a truncation error for the equation because each one of the finite-difference quotients has been taken from a truncated series. Considering equations (1) and (3) and looking at the truncation errors associated with the difference quotients we can write:

$$\frac{\partial u}{\partial t} - \alpha \frac{\partial^2 u}{\partial x^2} = \frac{u_i^{n+1} - u_i^n}{\Delta t} - \alpha \frac{u_{i+1}^n - 2u_i^n + u_{i-1}^n}{\left(\Delta x^2\right)}$$

$$+ \left[-\left(\frac{\partial^2 u}{\partial t^2}\right)_i^n \frac{(\Delta t)}{2} + \alpha \left(\frac{\partial^4 u}{\partial x^4}\right)_i^n \frac{(\Delta x)^2}{12} + ... \right] \qquad \qquad ...(3)$$

In equation (3), the terms in the square brackets represent truncation error for the complete equation. It is evident that the truncation error (TE) for this representation is $O\left[\Delta t, (\Delta x)^2\right]$.

With respect to equation (3), it can be said that as $\Delta x \to 0$ and $\Delta t \to 0$, the truncation error approaches zero. Hence, in the limiting case, the difference equation also approaches the original differential equation. Under such circumstances, the finite difference representation of the partial differential equation is said to be consistent.

Consistency

A finite difference representation of a partial differential equation (PDE) is said to be consistent if we can show that the difference between the PDE and its finite difference (FDE) representation vanishes as the mesh is refined, i.e:

$$\lim_{\text{mesh} \to 0} (PDE - FDE) = \lim_{\text{mesh} \to 0} (TE) = 0$$

A questionable scheme would be one for which the truncation error is $O\left(\Delta t / \Delta x\right)$ and not explicitly $O(\Delta t)$ or $O(\Delta x)$ or higher orders. In such cases the scheme would not be formally consistent unless the mesh were refined in a manner such that $(\Delta t / \Delta x) \to 0$.

Let us take equation (1) and the Dufort-Frankel (1953) differencing scheme.

The FDE is:

$$\frac{u_i^{n+1} - u_i^{n-1}}{2\Delta t} = \alpha \left[\frac{u_{i+1}^{n-1} - u_i^{n+1} - u_i^{n-1} + u_{i-1}^n}{(\Delta x^2)} \right] \qquad \ldots(4)$$

The above expression for truncation error is meaningful if $(\Delta t / \Delta x) \to 0$ together with $\Delta t \to 0$ and $\Delta x \to 0$. However, (Δt) and (Δx) may individually approach zero in such a way that $(\Delta t / \Delta x) = \beta$. Then if we reconstitute the PDE from FDE and TE, we shall obtain:

$$\lim_{\Delta t, \Delta x \to 0} (PDE - FDE) = \lim_{\text{mesh} \to 0} (TE) = -\alpha\beta^2 \frac{\partial^2 u}{\partial t^2}$$

And finally PDE becomes,

$$\frac{\partial u}{\partial t} + \alpha\beta^2 \frac{\partial^2 u}{\partial t^2} = \alpha \frac{\partial^2 u}{\partial x^2}$$

We started with a parabolic one and ended with a hyperbolic one. So, Dufort-Frankel scheme is not consistent for the 1D unsteady state heat conduction equation unless $(\Delta t / \Delta x) \to 0$ together with $\Delta t \to 0$ and $\Delta x \to 0$.

Stability

We are relying on the assumption that solving the difference equations gives a decent approximation to the solution of the underlying differential equations (actually the converse now, that the solution to the differential equation (2) gives a good indication of the solution to the difference equations (1)). Since it is exactly this assumption we are trying to prove, the reasoning is rather circular.

$$AE = -\tau. \qquad\qquad\qquad \text{...(1)}$$

$$e''(x) = -\tau(x) \quad \text{for} \quad 0 < x < 1 \qquad\qquad \text{...(2)}$$

Instead, let's look directly at the discrete system (1), which we will rewrite in the form:

$$A^h E^h = -\tau^h \qquad\qquad\qquad \text{...(3)}$$

Where, the superscript h indicates that we are on a grid with mesh spacing h. This serves as a reminder that these quantities change as we refine the grid. In particular, the matrix Ah is an m x m matrix with h=1(m+1) so that its dimension is growing as h →0.

Let (Ah)-1 be the inverse of this matrix. Then solving the system (3) gives:

$$E^h = -\left(A^h\right)^{-1}\tau^h$$

And taking norms gives:

$$\left\|E^h\right\| = \left\|\left(A^h\right)^{-1}\tau^h\right\|$$

$$\leq \left\|\left(A^h\right)^{-1}\right\|\left\|\tau^h\right\|.$$

We know that $\left\|\tau^h\right\| = O\left(h^2\right)$ and we are hoping the same will be true of $\left\|E^h\right\|$ It is clear what we need for this to be true. We need $\left\|\left(A^h\right)^{-1}\right\|$ to be bounded by some constant independent of h as h →0.

$$\left\|\left(A^h\right)^{-1}\right\| \leq C \text{ For all h sufficiently small.}$$

Then we will have:

$$\left\|E^h\right\| \leq C\left\|\tau^h\right\|$$

And so $\left\|E^h\right\|$ goes to zero at least as fast as $\left\|\tau^h\right\|$. This motivates the following definition of stability for linear BVPs.

Convergence

A method is said to be convergent if Combining the ideas introduced above we arrive at the conclusion that,

$$\text{Consistency + Stability} \rightarrow \text{Convergence.} \qquad \text{...(1)}$$

$$\left\|\left(A^h\right)^{-1}\right\| \leq C \text{ for all } h < h_o \qquad \text{...(2)}$$

$$\left\|\tau^h\right\| \rightarrow 0 \text{ as } h \rightarrow 0. \qquad \text{...(3)}$$

This is easily proved by using (2) and (3) to obtain the bound.

$$\left\|E^h\right\| \leq \left\|\left(A^h\right)^{-1}\right\| \left\|\tau^h\right\| \leq C \left\|\tau^h\right\| \rightarrow 0 \qquad \text{as} \qquad h \rightarrow 0. \qquad \text{...(4)}$$

Although this has been demonstrated only for the linear BVP, in fact most analyses of finite difference methods for differential equations follow this same two-tier approach, and the statement (1) is sometimes called the fundamental theorem of finite difference methods.

In fact, as our above analysis indicates, this can generally be strengthened to say that:

$$O(h^p) \text{ Local truncation error + Stability} \rightarrow O(h^p) \text{ global error} \quad \text{...(5)}$$

Consistency (and the order of accuracy) is usually the easy part to check. Verifying stability is the hard part. Even for the linear BVP just discussed it is not at all clear how to check the condition (2) since these matrices become larger as h $\rightarrow$ 0. For other problems it may not even be clear how to define stability in an appropriate way. As we will see, there are many definitions of "stability" for different types of problems.

The challenge in analyzing finite difference methods for new classes of problems often is to find an appropriate definition of "stability" that allows one to prove convergence using (1) while at the same time being sufficiently manageable that we can verify it holds for specific finite difference methods. For nonlinear PDEs this frequently must be tuned to each particular class of problems and relies on existing mathematical theory and techniques of analysis for this class of problems.

1.6 Time Integration: Finite Difference Methods in Solution of Steady and Unsteady Problem

Explicit and Implicit Time Integration Methods

$$\frac{u_i^{n+1} - u_i^n}{\Delta t} = \alpha \left[\frac{u_{i+1}^n - 2u_i^n + u_{i-1}^n}{\left(\Delta x^2\right)} \right] \qquad \ldots(1)$$

The solution of equation (1) takes the form of a "marching" procedure (or scheme) in steps of time. We know the dependent variable at all x at a time level from given initial conditions. Examining equation (1) we see that it contains one unknown, namely u_i^{n+1}. Thus, the dependent variable at time $(t + \Delta t)$ is obtained directly from the known values of u_{i+1}^n, u_i^n and u_{i-1}^n.

$$\frac{u_i^{n+1} - u_i^n}{\Delta t} = \alpha \left[\frac{u_{i+1}^n - 2u_i^n + u_{i-1}^n}{\left(\Delta x^2\right)} \right] \qquad \ldots(2)$$

This is a typical example of an explicit finite difference method.

Let us now attempt a different discretization of the original partial differential equation given by equation (1). Here we express the spatial differences on the right-hand side in terms of averages between n and $(n + 1)$ time levels,

$$\frac{u_i^{n+1} - u_i^n}{\Delta t} = \frac{\alpha}{2} \left[\frac{u_{i+1}^{n+1} + u_{i+1}^n - 2u_i^{n+1} - 2u_i^n + u_{i-1}^{n+1} + u_{i-1}^n}{\left(\Delta x^2\right)} \right] \qquad \ldots(3)$$

The differencing shown in equation (3) is known as the Crank-Nicolson implicit scheme. The unknown u_i^{n+1} is not only expressed in terms of the known quantities at time level n, but also in terms of unknown quantities at time level $(n+1)$.

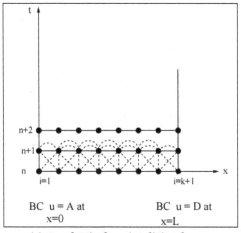

(a) Crank Nicolson implicit scheme.

Hence equation (3) at a given grid point i, cannot itself result in a solution of u_{i}^{n+1}. The equation (3) has to be written at all grid points, resulting in a system of algebraic equations from which the unknowns u_i^{n+1} for all i can be solved simultaneously.

This is a typical example of an implicit finite-different solution in shown in (Figure (a)). Since they deal with the solution of large systems of simultaneous linear algebraic equations, implicit methods usually require the handling of large matrices.

Generally, the following steps are followed in order to obtain a solution. The equation (3) can be rewritten as:

$$u_{i}^{n+1} - u_{i}^{n} = \frac{r}{2}\left[u_{i+1}^{n+1} + u_{i+1}^{n} - 2u_{i}^{n+1} - 2u_{i}^{n} + u_{i-1}^{n+1} + u_{i-1}^{n}\right] \qquad \ldots(4)$$

Where, $r = \alpha(\Delta t / \Delta x)^2$ or,

$$-r\,u_{i+1}^{n+1} + (2+2r)u_{i}^{n+1} - r\,u_{i+1}^{n+1} = r\,u_{i-1}^{n} + (2-2r)u_{i}^{n} + r\,u_{i+1}^{n}$$

Or $-u_{i+1}^{n+1} + \left(\dfrac{2+2r}{r}\right)u_{i}^{n+1} - u_{i+1}^{n+1} = u_{i-1}^{n} + \left(\dfrac{2-2r}{r}\right)u_{i}^{n} + u_{i+1}^{n}$ $\qquad \ldots(5)$

The equation (5) has to be applied at all grid points, i.e., from i = 1 to i = k + 1. A system of algebraic equations will result (Figure (a)):

at $\qquad$ i=2 $\qquad$ $-A + B(1)u_{2}^{n+1} - u_{3}^{n+1} = C(1)$

at $\qquad$ i = 3 $\qquad$ $-u_{2}^{n+1} + B(2)u_{3}^{n+1} - u_{4}^{n+1} = C(2)$

at $\qquad$ i = 4 $\qquad$ $-u_{3}^{n+1} + B(3)u_{4}^{n+1} - u_{5}^{n+1} = C(3)$

$\qquad\qquad\quad \vdots \qquad\qquad \vdots$

at $\qquad$ i = k $\qquad$ $-u_{k-1}^{n+1} + B(k-1)u_{k}^{n+1} - D = C(k-1)$

Finally the equations will be of the form:

$$\begin{bmatrix} B(1) & -1 & 0 & \cdots & 0 \\ -1 & B(2) & -1 & \cdots & 0 \\ 0 & -1 & B(3) & \cdots & 0 \\ \vdots & & & & \\ 0 & 0 & 0 & -1 & B(k-1) \end{bmatrix} \begin{bmatrix} u_{2}^{n+1} \\ u_{3}^{n+1} \\ u_{4}^{n+1} \\ \vdots \\ u_{k}^{n+1} \end{bmatrix} = \begin{bmatrix} (C(1)+A)^{n} \\ C(2)^{n} \\ C(3)^{n} \\ \vdots \\ (C(k-1)+D^{n}) \end{bmatrix} \qquad \ldots(6)$$

Here, we express the system of equations in the form of Ax = C:

Where,

$\qquad$ C - Right-hand side column vector (known)

A - Tridiagonal coefficient matrix (known)

x - Solution vector (to be determined)

Note that the boundary values at i=1 and i = k+1 are transferred to the known right-hand side.

For such a tridiagonal system, different solution procedures are available. In order to derive advantage of the zeros in the coefficient-matrix, the well-known Thomas algorithm (1949) can be used.

Explicit and Implicit Methods for Two-Dimensional Heat Conduction Equation

The two-dimensional conduction equation is given by:

$$\frac{\partial u}{\partial t} = \alpha \left(\frac{\partial^2 u}{\partial x^2} + \frac{\partial^2 u}{\partial y^2} \right) \qquad ...(7)$$

Here, the dependent variable, u (temperature) is a function of space (x, y) and time (t) and α is the thermal diffusivity. If we apply the simple explicit method to the heat conduction equation, the following algorithm results:

$$\frac{u_{i,j}^{n+1} - u_{i,j}^{n}}{\Delta t} = \alpha \left[\frac{u_{i+1,j}^{n} + 2u_{i,j}^{n} + u_{i-1,j}^{n}}{\left(\Delta x^2\right)} + \frac{u_{i,j+1}^{n} - 2u_{i,j}^{n} + u_{i,j-1}^{n}}{\left(\Delta y^2\right)} \right] \qquad ...(8)$$

When we apply the Crank-Nicolson scheme to the two-dimensional heat conduction equation, we obtain by:

$$\frac{u_{i,j}^{n+1} - u_{i,j}^{n}}{\Delta t} = \frac{\alpha}{2} \left(\delta_x^2 + \delta_y^2 \right) \left(u_{i,j}^{n+1} + u_{i,j}^{n} \right) \qquad ...(9)$$

Where, the central difference operators δ_x^2 and δ_y^2 in two different spatial directions are defined by:

$$\delta_x^2 \left[u_{i,j}^{n} \right] = \frac{u_{i+1,j}^{n} + 2u_{i,j}^{n} + u_{i-1,j}^{n}}{\left(\Delta x^2\right)}$$

$$\delta_y^2 \left[u_{i,j}^{n} \right] = \frac{u_{i,j+1}^{n} - 2u_{i,j}^{n} + u_{i,j-1}^{n}}{\left(\Delta y^2\right)} \qquad ...(10)$$

The resulting system of linear algebraic equations is not tridiagonal because of the five unknown's $u_{i,j}^{n+1}, u_{i+1,j}^{n+1}, u_{i-1,j}^{n+1}, u_{i,j+1}^{n+1}$, and $u_{i,j-1}^{n+1}$. In order to examine this further, let us re-write equation (9) as given by:

$$a\,u_{i,j-1}^{n+1} + b\,u_{i-1,j}^{n+1} + d\,u_{i,j}^{n+1} + b\,u_{i+1,j}^{n+1} + a\,u_{i,j+1}^{n+1} = c_{i,j}^{n} \qquad \ldots(11)$$

Where,

$$a = -\frac{\alpha\,\Delta t}{2(\Delta y)^2} = -\frac{1}{2}P_y$$

$$b = -\frac{\alpha\,\Delta t}{2(\Delta x)^2} = -\frac{1}{2}P_x$$

$$d = 1 + P_x + P_y$$

$$c_{i,j}^{n} = u_{i,j}^{n} + \frac{\alpha\,\Delta t}{2}\left(\delta_x^2 + \delta_y^2\right)u_{i,j}^{n}$$

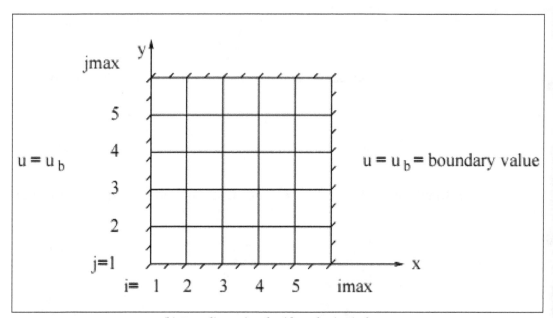

(b) Two-dimensional grid on the (x-y) plane.

The equation (11) can be applied to the two-dimensional (6 × 6) computational grid shown in the figure (b). A system of 16 linear algebraic equations have to be solved at (n + 1) time level, in order to get the temperature distribution inside the domain. The matrix equation will be as the following:

$$
\begin{bmatrix}
d & b & 0 & 0 & a & 0 & & & & & & & & & & & & 0 \\
b & d & b & & & a & & & & & & & & & & & & \\
0 & b & d & b & & & a & & & & & & & & & & & \\
0 & & b & d & b & & & a & & & & & & & & & & \\
a & & & 0 & d & b & & & a & & & & & & & & & \\
0 & a & & & b & d & b & & & a & & & & & & & & \\
& & a & 0 & & b & d & b & & & a & & & & & & & \\
& & & a & & & b & d & 0 & & & a & & & & & & \\
& & & & a & & & 0 & d & b & & & a & & & & & \\
& & & & & a & & & b & d & b & & & a & & & & \\
& & & & & & a & & & b & d & b & & & a & 0 & & \\
& & & & & & & a & & & b & d & 0 & & & a & & \\
& & & & & & & & a & & & 0 & d & b & & & 0 & \\
& & & & & & & & & a & & & b & d & b & & 0 & \\
& & & & & & & & & & a & & & b & d & b & 0 & \\
& & & & & & & & & & & a & & & b & d & b & \\
& & & & & & & & & & & & a & & & b & d &
\end{bmatrix}
\begin{bmatrix}
u_{2,2} \\ u_{3,2} \\ u_{4,2} \\ u_{5,2} \\ u_{2,3} \\ u_{3,3} \\ u_{4,3} \\ u_{5,3} \\ u_{2,4} \\ u_{3,4} \\ u_{4,4} \\ u_{5,4} \\ u_{2,5} \\ u_{3,5} \\ u_{4,5} \\ u_{5,5}
\end{bmatrix}
=
\begin{bmatrix}
c'''_{2,2} \\ c''_{3,2} \\ c''_{4,2} \\ c'''_{5,2} \\ c''_{2,3} \\ c_{3,3} \\ c_{4,3} \\ c''_{5,3} \\ c'_{2,4} \\ c_{3,4} \\ c_{4,4} \\ c''_{5,4} \\ c'''_{2,5} \\ c'_{3,5} \\ c'_{4,5} \\ c'''_{5,5}
\end{bmatrix}
\quad \ldots(12)
$$

Where,

$$c' = c - a\,u_b$$

$$c'' = c - b\,u_b$$

$$c''' = c - (a + b)u_b$$

The system of equations, described by equation (12) requires substantially more computer time as compared to a tridiagonal system. The equations of this type are usually solved by iterative methods. The quantity u_b is the boundary value.

ADI Method

The difficulties described in which occur when solving the two-dimensional equation by conventional algorithms, can be removed by alternating direction implicit (ADI) methods. The usual ADI method is a two-step scheme given by:

$$\frac{u_{i,j}^{n+1/2} - u_{i,j}^{n}}{\Delta t / 2} = \alpha\left(\delta_x^2 u_{i,j}^{n+1/2} + \delta_y^2\, u_{i,j}^{n}\right) \qquad \ldots(13)$$

and

$$\frac{u_{i,j}^{n+1} - u_{i,j}^{n+1/2}}{\Delta t / 2} = \alpha\left(\delta_x^2 u_{i,j}^{n+1/2} + \delta_y^2\, u_{i,j}^{n+1}\right) \qquad \ldots(14)$$

The effect of splitting the time step culminates in two sets of systems of linear algebraic equations. During step 1, we get the following:

$$\frac{u_{i,j}^{n+1/2} - u_{i,j}^n}{(\Delta t/2)} = \alpha\left[\left\{\frac{u_{i+1,j}^{n+1/2} - 2u_{i,j}^{n+1/2} + u_{i-1,j}^{n+1/2}}{(\Delta x^2)}\right\} + \left\{\frac{u_{i,j+1}^n - 2u_{i,j}^n + u_{i,j-1}^n}{(\Delta y^2)}\right\}\right]$$

or

$$\left[b\,u_{i-1,j} + (1-2b)\,u_{i,j} + b\,u_{i+1,j}\right]^{n+1/2} = u_{i,j}^n - a\left[u_{i,j+1} - 2u_{i,j} + u_{i,j-1}\right]^n$$

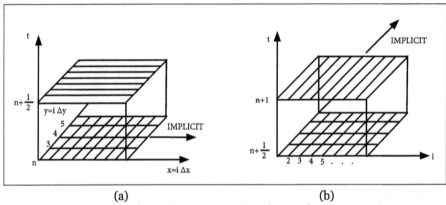

(a)　　　　　　　　　　　(b)

(1) Schematic representation of ADI scheme.

Now for each "j" rows (j = 2, 3...), we can formulate a tridiagonal matrix, for the varying i index and obtain the values from i = 2 to (imax − 1) at (n + 1/2) level the figure (1(a)). Similarly, in step-2, we get:

$$\frac{u_{i,j}^{n+1} - u_{i,j}^{n+1/2}}{(\Delta t/2)} = \alpha\left[\left\{\frac{u_{i+1,j}^{n+1/2} - 2u_{i,j}^{n+1/2} + u_{i-1,j}^{n+1/2}}{(\Delta x^2)}\right\} + \left\{\frac{u_{i,j+1}^{n+1} - 2u_{i,j}^{n+1} + u_{i,j-1}^{n+1}}{(\Delta y^2)}\right\}\right]$$

Or

$$\left[a\,u_{i,j-1} + (1-2a)\,u_{i,j} + a\,u_{i,j+1}\right]^{n+1} = b\left[u_{i+1,j} + -2u_{i,j} + u_{i-1,j}\right]^{n+1/2}$$

Now for each "i" rows (i = 2, 3....), we can formulate another tridiagonal matrix for the varying j index and obtain the values from j = 2 to (jmax − 1) at nth level shown in figure 1(b).

With a little more effort, it can be shown that the ADI method is also second- order accurate in time. If we use Taylor series expansion around $u_{i,j}^{n+1/2}$ on either direction, we shall obtain.

$$u_{i,j}^{n+1} = u_{i,j}^{n+1/2} + \left(\frac{\partial u}{\partial t}\right)\left(\frac{\Delta t}{2}\right) + \frac{1}{2!}\left(\frac{\partial^2 u}{\partial t^2}\right)\left(\frac{\Delta t}{2}\right)^2 + \frac{1}{3!}\left(\frac{\partial^3 u}{\partial t^3}\right)\left(\frac{\Delta t}{2}\right)^3 + \dots$$

and

$$u_{i,j}^{n} = u_{i,j}^{n+1/2} + \left(\frac{\partial u}{\partial t}\right)\left(\frac{\Delta t}{2}\right) + \frac{1}{2!}\left(\frac{\partial^2 u}{\partial t^2}\right)\left(\frac{\Delta t}{2}\right)^2 - \frac{1}{3!}\left(\frac{\partial^3 u}{\partial t^3}\right)\left(\frac{\Delta t}{2}\right)^3 + \ldots$$

Subtracting the latter from the former, one obtains,

$$u_{i,j}^{n+1} - u_{i,j}^{n} = \left(\frac{\partial u}{\partial t}\right)(\Delta t) + \frac{2}{3!}\left(\frac{\partial^3 u}{\partial t^3}\right)\left(\frac{\Delta t}{2}\right)^3 + \ldots$$

Or

$$\frac{\partial u}{\partial t} = \frac{u_{i,j}^{n+1} - u_{i,j}^{n}}{\Delta t} - \frac{1}{3!}\left(\frac{\partial^3 u}{\partial t^3}\right)\left(\frac{\Delta t}{2}\right)^2 + \ldots$$

The procedure above reveals that the ADI method is second-order accurate with a truncation error of $O\,[(\Delta t)^2,(\Delta x)^2,(\Delta y)^2]$.

The major advantages and disadvantages of explicit and implicit methods are summarized as follows:

Explicit

Advantage:

- The solution algorithm is simple to set up.

Disadvantage:

- For a given Δx, Δt must be less than a specific limit imposed by stability constraints. This requires many time steps to carry out the calculations over a given interval of t.

Implicit

Advantage:

- Stability can be maintained over much larger values of Δt. Fewer time steps are needed to carry out the calculations over a given interval.

Disadvantages:

- More involved procedure is needed for setting up the solution algorithm than that for explicit method.

- Since matrix manipulations are usually required at each time step, the computer time per time step is larger than that of the explicit approach.

- Since larger Δt can be taken, the truncation error is often large and the exact transients may not be captured accurately by the implicit scheme as compared to an explicit scheme.

Apparently finite-difference solutions seem to be straightforward. The overall procedure is to replace the partial derivatives in the governing equations with finite difference approximations and then finding out the numerical value of the dependent variables at each grid point.

Steady State Problem

In general we expect the temperature distribution to change with time. However, if $\psi(x,t)$, $\alpha(t)$ and $\beta(t)$ are all time-independent, then we might expect the solution to eventually reach a steady-state solution $u(x)$ which then remains essentially unchanged at later times.

$$u_t(x,t) = \kappa u_{xx}(x,t) + \psi(x,t) \qquad \text{... 1} \qquad \qquad \text{...(1)}$$

Typically there will be an initial transient time, as the initial data $uo(x)$ approaches $u(x)$ (unless $uo(x) \equiv u(x)$), but if we are only interested in computing the steady state solution itself, then we can set ut = 0 in equation (1) and obtain an ordinary differential equation in x to solve for $u(x)$:

$$u^n(x) = f(x) \qquad \qquad \text{...(2)}$$

Where we introduce $f(x) = -\psi(x)/\kappa$ to avoid minus signs below. This is a second order ODE and from basic theory we expect to need two boundary conditions in order to specify a unique solution. In our case we have the boundary conditions:

$$u(a) = \alpha, \ u(b) = \beta \qquad \qquad \text{...(3)}$$

The equation (2), (3) is called a two-point boundary value problem since one condition is specified at each of the two endpoints of the interval where the solution is desired.

If instead we had 2 data values specified at the same point, say $u(a) = \alpha$, $u'(a) = \sigma$ and we want to find the solution for t ≥ a, then we would have an initial value problem instead.

One approach to computing a numerical solution to a steady state problem is to choose some initial data and March forward in time using a numerical method for the time-dependent partial differential equation (1), on the solution of parabolic equations.

However, this is typically not an efficient way to compute the steady-state solution if this is all we want. Instead we can discretized and solve the two-point boundary value problem given by (2) and (3) directly.

Unsteady State Problem

For analyzing the equations for fluid flow problems, it is convenient to consider the case of a second-order differential equation given in the general form as:

$$A\frac{\partial^2 \phi}{\partial x^2}+B\frac{\partial^2 \phi}{\partial x \partial y}+C\frac{\partial^2 \phi}{\partial y^2}+D\frac{\partial \phi}{\partial x}+E\frac{\partial \phi}{\partial y}+F\phi=G(x,y) \qquad ...(1)$$

In the coefficients A, B, C, D, E and F are either constants or functions of only (x, y) (do not contain φ or its derivatives), it is said to be a linear equation, otherwise it is a non-linear equation.

An important subclass of non-linear equations is quasi-linear equations. In this case, the coefficients may contain φ or its first derivative but not the second (highest) derivative. If G = 0, the aforesaid equation is homogeneous, otherwise it is non-homogeneous.

Again for the above mentioned equation:

if $B^2 - 4AC = 0$, the equation is parabolic.

if $B^2 - 4AC < 0$, the equation is elliptic.

if $B^2 - 4AC > 0$, the equation is hyperbolic.

The unsteady Navier-Stokes equations are elliptic in space and parabolic in time. At steady-state, the Navier-Stokes equations are elliptic. In elliptic problems, the boundary conditions must be applied on all confining surfaces. These are boundary value problems. A physical problem may be steady or unsteady.

The Laplace equations and the Poisson equations are generally associated with the steady-state problems. These are elliptic equations and can be written respectively as:

$$\frac{\partial^2 \phi}{\partial x^2}+\frac{\partial^2 \phi}{\partial y^2}=0 \qquad ...(2)$$

$$\frac{\partial^2 \phi}{\partial x^2}+\frac{\partial^2 \phi}{\partial y^2}+S=0 \qquad ...(3)$$

The velocity potential in steady, in viscid, incompressible and irrotational flows satisfies the Laplace equation. The temperature distribution for steady-state, constant-property, two-dimensional conduction satisfies the Laplace equation if no volumetric heat source is present in the domain of interest and the Poisson equation if a volumetric heat source is present.

The parabolic equation in conduction heat transfer is of the form:

$$\frac{\partial \phi}{\partial t}=B\frac{\partial^2 \phi}{\partial x^2} \qquad ...(4)$$

The one-dimensional unsteady conduction problem is governed by this equation when t and x are identified as the time and space variables, respectively φ denotes the temperature and B is the thermal diffusivity.

The boundary conditions at the two ends and an initial condition are needed to solve such equations. The unsteady conduction problem in two-dimensions is governed by an equation of the form,

$$\frac{\partial \phi}{\partial t} = B\left(\frac{\partial^2 \phi}{\partial x^2} + \frac{\partial^2 \phi}{\partial y^2}\right) + S \qquad \ldots(5)$$

Where, t denotes the time variable and a source term S is included. By comparing the highest derivatives in any two of the independent variables, with the help of the conditions given earlier, it can be concluded that equation (5) is parabolic in time and elliptic in space. An initial condition and two conditions for the extreme ends in each special coordinate are required to solve this equation.

Fluid flow problems generally have nonlinear terms due to the inertia or acceleration component in the momentum equation. These terms are called advection terms.

The energy equation has nearly similar terms, usually called the convection terms, which involve the motion of the flow field. For unsteady two-dimensional problems, the appropriate equations can be represented as:

$$\frac{\partial \phi}{\partial t} + u\frac{\partial \phi}{\partial x} + v\frac{\partial \phi}{\partial y} = B\left(\frac{\partial^2 \phi}{\partial x^2} + \frac{\partial^2 \phi}{\partial y^2}\right) + S \qquad \ldots(6)$$

Where,

φ - Velocity, temperature or some other transported property,

B - The diffusivity for momentum or heat,

u and v - Velocity components,

S - Source term.

The pressure gradients in the momentum or the volumetric heating in the energy equation can be appropriately substituted in S. Equation (6) is parabolic in time and elliptic in space. However, for very high-speed flows, the terms on the left side dominate, the second-order terms on the right hand side become trivial and the equation becomes hyperbolic in time and space.

Boundary and Initial Conditions

In addition to the governing differential equations, the formulation of the problem requires a complete specification of the geometry of interest and appropriate boundary conditions.

An arbitrary domain and bounding surfaces are shown in the figure (a). The conservation equations are to be applied within the domain. The number of boundary conditions required is generally determined by the order of the highest derivatives appearing in each independent variable in the governing differential equations.

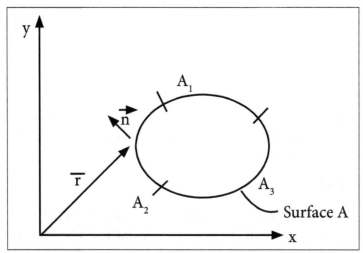

(a) Schematic sketch of an arbitrary domain.

The unsteady problems governed by a first derivative in time will require initial condition in order to carry out the time integration. The diffusion terms require two spatial boundary conditions for each coordinate in which a second derivative appears.

The spatial boundary conditions in flow and heat transfer problems are of three general types. They may be stated:

$$\phi = \phi_1(r) \in A_1 \qquad \ldots 1 \qquad\qquad \ldots(1)$$

$$\frac{\partial \phi}{\partial n} = \phi_2(r) \in A_2 \qquad\qquad \ldots(2)$$

$$a(r)\phi + b(r)\frac{\partial \phi}{\partial n} = \phi_3(r) \in A_3 \qquad\qquad \ldots(3)$$

Where A_1, A_2 and A_3 denote three separate zones on the bounding surface in the figure(a). The boundary conditions on φ in equations (1) to (3) are usually referred to as Dirchlet, Neumann and mixed boundary conditions, respectively.

The boundary conditions are linear in the dependent variable φ. In equations (1) and (2), $\vec{r} = \vec{r}(x,y)$ is a vector denoting position on the boundary.

$\dfrac{\partial}{\partial n}$ is the directional derivative normal to the boundary and φ_1, φ_2, φ_3, a and b are arbitrary functions. The normal derivative may be expressed as:

$$\frac{\partial \phi}{\partial n} = \vec{n} \cdot \nabla \phi$$

$$= \left(n_x \hat{i} + n_y \hat{j} \right) \cdot \left(\frac{\partial \phi}{\partial x} \hat{i} + \frac{\partial \phi}{\partial y} \hat{j} \right) \qquad \qquad ...(4)$$

$$= n_x \frac{\partial \phi}{\partial x} + n_y \frac{\partial \phi}{\partial y}$$

Here, $\vec{n}$ is the unit vector normal to the boundary, ∇ is the nabla operator, [•] denotes the dot product, (n_x, n_y) are the direction-cosine components of $\vec{n}$ and $\left(\hat{i}, \hat{j} \right)$ are the unit vectors aligned with the (x, y) coordinates.

Iterative Methods

These methods are used to solve a special linear equations in which each equation must possess one large coefficient and the large coefficient must be attached to a different unknown in that equation.

Further in each equation, the absolute value of the large coefficient of the unknown is greater than the sum of the absolute values of the other coefficients of the other unknowns. Such type of simultaneous linear equations can be solved by the following iterative methods:

- Gauss-Jacobi method

- Gauss-seidel method

$$a_1 x + b_1 y + c_1 z = d_1$$

$$a_2 x + b_2 y + c_2 z = d_2$$

$$a_3 x + b_3 y + c_3 z = d_3$$

i.e., the co-efficient matrix $A = \begin{bmatrix} \langle a_1 \rangle & b_1 & c_1 \\ a_2 & \langle b_2 \rangle & c_2 \\ a_3 & b_3 & \langle c_3 \rangle \end{bmatrix}$ is diagonally dominant. Solving the

given system for x, y, z (whose diagonals are the largest values), we have:

$$x = \frac{1}{a_1} \left[d_1 - b_1 y - c_1 z \right]$$

$$y = \frac{1}{b_2}[d_2 - a_2 x - c_2 z]$$

$$z = \frac{1}{c_3}[d_3 - a_3 x - b_3 y]$$

Gauss-Jacobi Method

If the rth iterates are x(r), y(r), z (r) then the iteration scheme for this method is given by:

$$x^{(r+1)} = \frac{1}{a_1}\left[d_1 - b_1 y^{(r)} - c_1 z^{(r)}\right]$$

$$y^{(r+1)} = \frac{1}{b_2}\left[d_2 - a_2 x^{(r)} - c_2 z^{(r)}\right]$$

$$z^{(r+1)} = \frac{1}{c_3}\left[d_3 - a_3 x^{(r)} - b_3 y^{(r)}\right]$$

The iteration is stopped when the values x, y, z start repeating with the desired degree of accuracy.

Gauss-Seidel Method

This method is only a refinement of Gauss-Jacobi method. In this method, once a new value for a unknown is found, it is used immediately for computing the new values of the unknowns.

If the rth iterates are $x^{(r+1)}$, $y^{(r+1)}$, $z^{(r+1)}$, then the iteration scheme for this method is given by:

$$x^{(r+1)} = \frac{1}{a_1}\left[d_1 - b_1 y^{(r)} - c_1 z^{(r)}\right]$$

$$y^{(r+1)} = \frac{1}{b_2}\left[d_2 - a_2 x^{(r+1)} - c_2 z^{(r)}\right]$$

$$z^{(r+1)} = \frac{1}{c_3}\left[d_3 - a_3 x^{(r+1)} - b_3 y^{(r+1)}\right]$$

Hence finding the values of the unknowns, we use the latest available values on the R.H.S. The process of iteration is continued until the convergence is obtained with desired accuracy.

FEM for Two and Three Dimensional Solids

2.1   The Continuum

Continuum Structures

Structures such as plates, thin shells, thick shells, solids, which do not have distinctly identifiable members, can be modeled by an arbitrary number of elements of different shapes viz., triangles and quadrilaterals in 2-D structures and tetrahedron and brick elements in 3-D structures. These are called continuum structures.

In these structures, adjacent elements have a common boundary surface. The finite element model represents true situation only when displacements and their significant derivate of adjacent elements are same along their common boundary.

Idealization of a Continuum

A continuum may be discretized in different ways depending upon the geometrical configuration of the domain. Figure (a) shows the various ways of idealizing a continuum based on the geometry.

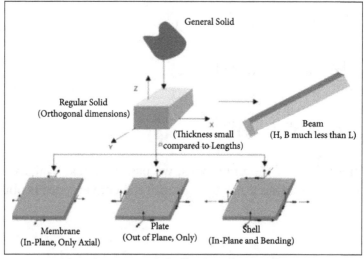

(a) Various ways of Idealization of a Continuum.

Discretization of Technique

The need of finite element analysis arises when the structural system in terms of its either geometry, material properties, boundary conditions or loadings is complex in nature.

For such case, the whole structure needs to be subdivided into smaller elements. The whole structure is then analyzed by the assemblage of all elements representing the complete structure including its all properties.

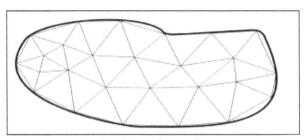

(a) Triangular mesh.

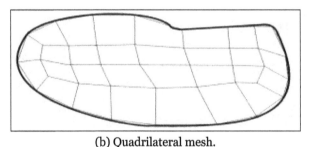

(b) Quadrilateral mesh.

(1) Discretization of a continuum.

The subdivision process is an important task in finite element analysis and requires some skill and knowledge. In this procedure, first, the number, shape, size and configuration of elements have to be decided in such a manner that the real structure is simulated as closely as possible.

The discretization is to be in such that the results converge to the true solution. However, too fine mesh will lead to extra computational effort. Figure (1) shows a finite element mesh of a continuum using triangular and quadrilateral elements.

The assemblage of triangular elements in this case shows better representation of the continuum. The discretization process also shows that the more accurate representation is possible if the body is further subdivided into some finer mesh.

2.1.1 Equations of Equilibrium

Considering all the forces acting, we can write equations of equilibrium for the element as:

$$\sum F_x = 0$$

$$\sigma_x^+ dy\,dz - \sigma_x\,dy\,dz + \tau_{yx}^+\,dx\,dz - \tau_{yx}\,dx\,dz + \tau_{zx}^+\,dx\,dy + X\,dx\,dy\,dz = 0$$

i.e. $\left(\sigma_x + \dfrac{\partial \sigma_x}{\partial x}dx\right)dy\,dz - \sigma_x\,dy\,dz + \left(\tau_{yx} + \dfrac{\partial \tau_{yx}}{\partial y}dy\right)dx\,dz - \tau_{yx}\,dx\,dz$

$$+\left(\tau_{zx} + \dfrac{\partial \tau_{zx}}{\partial z}dz\right)dy\,dx - \tau_{zx}\,dx\,dy + X\,dx\,dy\,dz = 0$$

Simplifying and then dividing throughout by dx dy dz, we get:

$$\frac{\partial \sigma_x}{\partial x} + \frac{\partial \tau_{yx}}{\partial y} + \frac{\partial \tau_{zx}}{\partial z} + X = 0$$

Similarly $\sum F_y = 0$ and $\sum F_z = 0$ equilibrium conditions gives:

$$\frac{\partial \tau_{xy}}{\partial x} + \frac{\partial \sigma_y}{\partial y} + \frac{\partial \tau_{zx}}{\partial z} + Y = 0$$

and

$$\frac{\partial \tau_{xy}}{\partial x} + \frac{\partial \sigma_{yz}}{\partial y} + \frac{\partial \tau_y}{\partial z} + Z = 0$$

Now, $\sum$ moment about x-axis=0 through the centroid of the element gives:

$$\tau_{yz}^+\,dx\,dz\frac{dy}{2} + \tau_{yz}\,dx\,dz\frac{dy}{2} - \left[\tau_{zy}^+\,dx\,dz\frac{dy}{2} + \tau_{zy}\,dx\,dz\frac{dy}{2}\right] = 0$$

i.e. $\left(\tau_{yz} + \dfrac{\partial \tau_{yz}}{\partial y}dy\right)dx\,dy\dfrac{dz}{2} + \tau_{yz}\,dx\,dy\dfrac{dy}{2}$

$$-\left[\left(\tau_{zy} + \frac{\partial \tau_{yz}}{\partial y}dz\right)dx\,dy\frac{dy}{2} + \tau_{zy}\,dx\,dz\frac{dz}{2}\right] = 0$$

Neglecting the small quantity of higher (4th) order and dividing throughout by dx dy dz, we get:

$$\tau_{yz} = \tau_{zy}$$

Similarly the moment equilibrium conditions about y-axis and z-axis result into:

$$\tau_{xz} = \tau_{zx}$$

and

$$\tau_{xy} = \tau_{yx}$$

Thus, the stress vector is given by:

$$|\sigma|^x = \begin{bmatrix} \sigma_x & \sigma_y & \sigma_z & \tau_{xy} & \tau_{yz} & \tau_{zx} \end{bmatrix}$$

and the equations of equilibrium are:

$$\frac{\partial \sigma_x}{\partial x} + \frac{\partial \tau_{xy}}{\partial y} + \frac{\partial \tau_{xz}}{\partial z} + X = 0$$

$$\frac{\partial \tau_{xy}}{\partial x} + \frac{\partial \sigma_y}{\partial y} + \frac{\partial \tau_{yz}}{\partial z} + Y = 0$$

and

$$\frac{\partial \tau_{xz}}{\partial x} + \frac{\partial \tau_{yz}}{\partial y} + \frac{\partial \sigma_z}{\partial z} + Z = 0$$

And note that

$$\tau_{xy} = \tau_{yx}, \tau_{yz} = \tau_{zy} \text{ and } \tau_{xz} = \tau_{zx}$$

2.1.2 Boundary Conditions

From the following procedure the effect of boundary condition in the stiffness matrix for the finite element analysis can be obtained. The solution cannot be obtained unless support conditions are included in the stiffness matrix.

This is because, if all the nodes of the structure are included in displacement vector, the stiffness matrix becomes singular and cannot be solved if the structure is not supported amply and it cannot resist the applied loads. A solution cannot be achieved until the boundary conditions i.e., the known displacements are introduced.

In finite element analysis, the partitioning of the global matrix is carried out in a systematic way for the hand calculation as well as for the development of computer codes.

In partitioning, normally the equilibrium equations can be partitioned by rearranging corresponding rows and columns, so that prescribed displacements are grouped together. For example, let consider the equation of equilibrium is expressed in compact form as:

$$\{F\} = [K] \{d\} \qquad\qquad ...(1)$$

Where,

$[K]$ - The global stiffness matrix.

$\{d\}$ - The displacement vector consisting of global degrees of freedom.

$\{F\}$ - The load vector corresponding to degrees of freedom.

By the method of partitioning the above equation can be partitioned in the following manner:

$$\begin{Bmatrix} \{F_\alpha\} \\ \{F_\beta\} \end{Bmatrix} = \begin{bmatrix} [K_{\alpha\alpha}] & [K_{\alpha\beta}] \\ [K_{\beta\alpha}] & [K_{\beta\beta}] \end{bmatrix} \begin{Bmatrix} \{d_\alpha\} \\ \{d_\beta\} \end{Bmatrix} \qquad \ldots(2)$$

Where, subscripts α refers to the displacements free to move and β refers to the prescribed support displacements. As the prescribed displacements $\{d_\beta\}$ are known, equation (2) may be written in expanded form as:

$$\{F_\alpha\} = [K_{\alpha\alpha}]\{d_\alpha\} + [K_{\alpha\beta}]\{d_\beta\} \qquad \ldots(3)$$

Thus it is possible to obtain the free displacement of the structure $\{d_\alpha\}$ as:

$$\{d_\alpha\} = [K_{\alpha\alpha}]-1\{\{F_\alpha\} - [K_{\alpha\beta}]\{d_\beta\}\} \qquad \ldots(4)$$

If the displacements at supports $\{d_\beta\}$ are zero, then the above equation can be simplified to the following expression:

$$\{d_\alpha\} = [K_{\alpha\alpha}]-1\{F_\alpha\} \qquad \ldots(5)$$

Thus, by rearranging assembled matrix, the portion corresponding to the unknown displacements in equation (4) can be taken out for the solution purpose. This is possible as the known displacements $\{d_\beta\}$ are restrained, i.e., displacements are zero.

If the support has some known displacements, then equation (4) can be used to find the solution. If the few supports of the structures yield, then the above method may be modified by partitioning the stiffness matrix into three parts as shown below:

$$\begin{Bmatrix} \{F_\alpha\} \\ \{F_\beta\} \\ \{F_\gamma\} \end{Bmatrix} = \begin{bmatrix} [K_{\alpha\alpha}] & [K_{\alpha\beta}] & [K_{\alpha\gamma}] \\ [K_{\beta\alpha}] & [K_{\beta\beta}] & [K_{\beta\gamma}] \\ [K_{\gamma\alpha}] & [K_{\gamma\beta}] & [K_{\gamma\gamma}] \end{bmatrix} \begin{Bmatrix} \{d_\alpha\} \\ \{d_\beta\} \\ \{d_\gamma\} \end{Bmatrix} \qquad \ldots(6)$$

Here, α refers to unknown displacement, β refers to known displacement ($\neq 0$) and γ

refers to zero displacement. Thus, the above equation can be separated and solved independently to find required unknown results as shown below:

$$\{F_\alpha\} = [K_{\alpha\alpha}]\{d_\alpha\} + [K_{\alpha\beta}]\{d_\beta\} + [K_{\alpha\gamma}]\{d_\gamma\}$$

Or

$$[K_{\alpha\alpha}]\{d_\alpha\} = \{F_\alpha\} - [K_{\alpha\beta}]\{d_\beta\} \text{ as } \{d_\gamma\} = \{0\}$$

Thus,

$$\{d_\alpha\} = [K_{\alpha\alpha}]^{-1}\{\{F_\alpha\} - [K_{\alpha\beta}]\{d_\beta\}\} \qquad \qquad ...(7)$$

For computer programming, several techniques are available for handling boundary conditions. One of the approaches is to make the diagonal element of stiffness matrix corresponding to zero displacement as unity and corresponding all off-diagonal elements as zero.

For example, let consider a 3x3 stiffness matrix with following force-displacement relationship:

$$\begin{Bmatrix} F_1 \\ F_2 \\ F_3 \end{Bmatrix} = \begin{bmatrix} k_{11} & k_{12} & k_{13} \\ k_{21} & k_{22} & k_{23} \\ k_{31} & k_{32} & k_{33} \end{bmatrix} \begin{Bmatrix} d_1 \\ d_2 \\ d_3 \end{Bmatrix} \qquad \qquad ...(8)$$

Now, if the third node has zero displacement (i.e., $d_3 = 0$) then the matrix will be modified as follows to incorporate the boundary condition:

$$\begin{Bmatrix} F_1 \\ F_2 \\ 0 \end{Bmatrix} = \begin{bmatrix} k_{11} & k_{12} & 0 \\ k_{21} & k_{22} & 0 \\ 0 & 0 & 1 \end{bmatrix} \begin{Bmatrix} d_1 \\ d_2 \\ d_3 \end{Bmatrix} \qquad \qquad ...(9)$$

Thus, while inverting whole matrix, d_3 will become zero automatically.

To incorporate known support displacement in computer programming following procedure may be adopted. Considering the displacement d_2 has known value of δ, 1st row of equation (8) can be written as:

$$F_1 = k_{11} \times d_1 + k_{12} \times d_2 + k_{13} \times d_3 \qquad \qquad ...(10)$$

Or

$$F_1 - k_{12} \times \delta = k_{11} \times d_1 + k_{13} \times d_3 \qquad \qquad ...(11)$$

Now, the 2nd row of equation (8) has to become:

$$\{\delta\} = \{d_2\} \quad \dots 12 \qquad \dots(12)$$

Similarly 3rd row will be:

$$F_3 - k_{32} \times \delta = k_{31} \times d_1 + k_{33} \times d_3 \qquad \dots(13)$$

Thus above three equations can be written in a combined form as:

$$\begin{Bmatrix} F_1 - k_{12}\delta \\ \delta \\ F_3 - k_{32}\delta \end{Bmatrix} = \begin{bmatrix} k_{11} & 0 & k_{13} \\ 0 & 1 & 0 \\ k_{31} & 0 & k_{33} \end{bmatrix} \begin{Bmatrix} d_1 \\ d_2 \\ d_3 \end{Bmatrix} \qquad \dots(14)$$

Another approach may also be followed to take care the known restrained displacements by assigning a higher value δ (say δ=1020) in the diagonal element corresponding to that displacement.

$$\begin{Bmatrix} F_1 \\ \delta \times 10^{20} \times k_{22} \\ F_3 \end{Bmatrix} = \begin{bmatrix} k_{11} & k_{12} & k_{13} \\ k_{21} & k_{22} \times 10^{20} & k_{23} \\ k_{31} & k_{32} & k_{33} \end{bmatrix} \begin{Bmatrix} d_1 \\ d_2 \\ d_3 \end{Bmatrix} \qquad \dots(15)$$

$$\therefore \delta \times 10^{20} \times k_{22} = k_{21}d_1 + k_{22} \times 10^{20} \times d_2 + k_{23} \times d_3$$

As d_3 is corresponding to zero displacement, the above equation can be simplified to the following:

$$\therefore \delta \times 10^{20} \times k_{22} = k_{21}d_1 + k_{22} \times 10^{20} \times d_2$$

$$\text{or } \delta \times 10^{20} \times k_{22} = k_{22} \times 10^{20} \times d^2$$

$$d_2 = \delta \rightarrow \text{known displacement is ensured.}$$

If the overall stiffness matrix is to be formed in half band form then the numbering of nodes should be such that the bandwidth is minimum. For this the labels are put in a systematic manner irrespective of whether the joint displacements are unknowns or restraints.

However, if the unknown displacements are labeled first then the matrix operations can be restricted up to unknown displacement labels and beyond that the overall stiffness matrix may be ignored.

2.2 Strain-Displacement Relations and Stress-Strain Relations

For two dimensional:

Strain

Corresponding to the six stress components, the state of strain at a point may be divided into six strain components as shown below:

$$\{\varepsilon\}^T = \left[\varepsilon_x \ \varepsilon_y \ \varepsilon_z \ \gamma_{xy} \ \gamma_{yz} \ \gamma_{yx} \right]$$

Strain-Displacement Equation

Taking displacement components in x, y, z directions as u, v and w respectively, the relations among components of strains and components of displacements are:

$$\varepsilon_x = \frac{\partial u}{\partial x} + \frac{1}{2}\left[\left(\frac{\partial u}{\partial x}\right)^2 + \left(\frac{\partial v}{\partial x}\right)^2 \left(\frac{\partial w}{\partial x}\right)^2\right]$$

$$\varepsilon_y = \frac{\partial v}{\partial y} + \frac{1}{2}\left[\left(\frac{\partial u}{\partial y}\right)^2 + \left(\frac{\partial v}{\partial y}\right)^2 \left(\frac{\partial w}{\partial y}\right)^2\right]$$

$$\varepsilon_z = \frac{\partial w}{\partial z} + \frac{1}{2}\left[\left(\frac{\partial u}{\partial z}\right)^2 + \left(\frac{\partial v}{\partial z}\right)^2 \left(\frac{\partial w}{\partial z}\right)^2\right]$$

$$\gamma_{xy} = \frac{\partial v}{\partial z} + \frac{\partial u}{\partial y} + \frac{\partial u}{\partial x}\cdot\frac{\partial u}{\partial y} + \frac{\partial v}{\partial x}\cdot\frac{\partial v}{\partial y} + \frac{\partial w}{\partial x}\cdot\frac{\partial w}{\partial y}$$

$$\gamma_{yz} = \frac{\partial w}{\partial y} + \frac{\partial v}{\partial z} + \frac{\partial u}{\partial y}\cdot\frac{\partial u}{\partial z} + \frac{\partial v}{\partial y}\cdot\frac{\partial v}{\partial z} + \frac{\partial w}{\partial y}\cdot\frac{\partial w}{\partial z}$$

$$\gamma_{xz} = \frac{\partial u}{\partial z} + \frac{\partial w}{\partial x} + \frac{\partial u}{\partial x}\cdot\frac{\partial u}{\partial z} + \frac{\partial v}{\partial x}\cdot\frac{\partial v}{\partial z} + \frac{\partial w}{\partial x}\cdot\frac{\partial w}{\partial z}$$

The strains are expressed up to the accuracy of second order changes in displacements. These equations may be simplified to the first order accuracy only by dropping the second order changes. Then linear strain displacement relation is given by:

$$\varepsilon_x = \frac{\partial u}{\partial x} \quad \gamma_{xy} = \frac{\partial u}{\partial x} + \frac{\partial v}{\partial y}$$

$$\varepsilon_y = \frac{\partial v}{\partial y} \quad \gamma_{yz} = \frac{\partial w}{\partial y} + \frac{\partial v}{\partial z}$$

$$\varepsilon_z = \frac{\partial w}{\partial z} \quad \gamma_{xz} = \frac{\partial w}{\partial x} + \frac{\partial u}{\partial z}$$

Linear Constitutive Equations

The constitutive law expresses the relationship among stresses and strains. In theory of elasticity, usually it is considered as linear. In one dimensional stress analysis, stress is proportional to strain and the constant of proportionality is known as Young's modulus. It is very well known as Hooke's law. The similar relation is expressed among the six components of stresses and strains and is called "Generalized Hooke's Law". This may be stated as:

$$
\begin{Bmatrix} \sigma_x \\ \sigma_y \\ \sigma_z \\ \tau_{xy} \\ \tau_{yz} \\ \tau_{xz} \end{Bmatrix} =
\begin{bmatrix}
D_{11} & D_{12} & D_{13} & D_{14} & D_{15} & D_{16} \\
D_{21} & D_{22} & D_{23} & D_{24} & D_{25} & D_{26} \\
D_{31} & D_{32} & D_{33} & D_{34} & D_{35} & D_{36} \\
D_{41} & D_{42} & D_{43} & D_{44} & D_{45} & D_{46} \\
D_{51} & D_{52} & D_{53} & D_{54} & D_{55} & D_{56} \\
D_{61} & D_{62} & D_{63} & D_{64} & D_{65} & D_{66}
\end{bmatrix}
\begin{Bmatrix} \varepsilon_x \\ \varepsilon_y \\ \varepsilon_z \\ \gamma_{xy} \\ \gamma_{yz} \\ \gamma_{xz} \end{Bmatrix}
$$

Or in matrix form,

$$\{\sigma\} = [D]\{\varepsilon\},$$

Where, D is a 6 × 6 matrix of constants of elasticity to be determined by experimental investigations for each material. As D is symmetric matrix $\left[D_{ij} = D_{ji}\right]$, there are 21 material properties for linear elastic Anisotropic Materials.

Certain materials exhibit symmetry with respect to planes within the body. Such materials are called Orthotropic materials. Hence for Orthotropic materials, the number of material constants reduces to 9 as show below:

$$
\begin{Bmatrix} \sigma_x \\ \sigma_y \\ \sigma_z \\ \tau_{xy} \\ \tau_{yz} \\ \tau_{xz} \end{Bmatrix} =
\begin{Bmatrix}
D_{11} & D_{12} & D_{13} & 0 & 0 & 0 \\
 & D_{22} & D_{23} & 0 & 0 & 0 \\
 & & D_{33} & 0 & 0 & 0 \\
 & \text{Sym} & & D_{44} & 0 & 0 \\
 & & & & D_{55} & 0 \\
 & & & & & D_{66}
\end{Bmatrix}
\begin{Bmatrix} \varepsilon_x \\ \varepsilon_y \\ \varepsilon_z \\ \gamma_{xy} \\ \gamma_{yz} \\ \gamma_{xz} \end{Bmatrix}
$$

Using the Young's Modulus and Poisons ratio terms the above relation may be expressed as:

$$\varepsilon_x = \frac{\sigma_x}{E_x} - \mu_{yx}\frac{\sigma_y}{E_y} - \mu_{zx}\frac{\sigma_z}{E_z}$$

$$\varepsilon_y = -\mu_{xy}\frac{\sigma_x}{E_x} + \frac{\sigma_y}{E_y} - \mu_{zy}\frac{\sigma_z}{E_z}$$

$$\varepsilon_z = -\mu_{xz}\frac{\sigma_x}{E_x} - \mu_{yz}\frac{\sigma_y}{E_y} + \frac{\sigma_z}{E_z}$$

$$\gamma_{xy} = \frac{\tau_{xy}}{G_{xy}}, \gamma_{yz} = \frac{\tau_{yz}}{G_{yz}}, \gamma_{zx} = \frac{\tau_{zx}}{G_{zx}}$$

$$\frac{E_x}{\mu_{xy}} = \frac{E_y}{\mu_{yx}}, \frac{E_y}{\mu_{yz}} = \frac{E_z}{\mu_{zy}}, \frac{E_z}{\mu_{zx}} = \frac{E_x}{\mu_{xz}}$$

For Isotropic Materials the above set of equations are further simplified. An isotropic material is the one that has same material property in all directions. In other word for isotropic materials:

$$E_x = E_y = E_z \quad \text{say } E$$

$$\mu_{xy} = \mu_{yx} = \mu_{yz} = \mu_{zy} = \mu_{xz} = \mu_{zx} \quad \text{say } \mu$$

Hence for a three dimensional problem, the strain stress relation for isotropic material is:

$$\begin{Bmatrix} \varepsilon_x \\ \varepsilon_y \\ \varepsilon_z \\ \gamma_{xy} \\ \gamma_{yz} \\ \gamma_{xz} \end{Bmatrix} = \begin{bmatrix} \frac{1}{E} & -\frac{\mu}{E} & -\frac{\mu}{E} & 0 & 0 & 0 \\ & \frac{1}{E} & -\frac{\mu}{E} & 0 & 0 & 0 \\ & & \frac{1}{E} & 0 & 0 & 0 \\ & & & \frac{1-\mu}{2} & 0 & 0 \\ & & & & \frac{1-\mu}{2} & 0 \\ & & & & & \frac{1-\mu}{2} \end{bmatrix} \begin{Bmatrix} \sigma_x \\ \sigma_y \\ \sigma_z \\ \tau_{xy} \\ \tau_{yz} \\ \tau_{xz} \end{Bmatrix}$$

Since, $G = \dfrac{E}{2(1-\mu)}$ and stress-strain relation is given by:

$$\begin{Bmatrix} \sigma_x \\ \sigma_y \\ \sigma_z \\ \tau_{xy} \\ \tau_{yz} \\ \tau_{xz} \end{Bmatrix} = \dfrac{E}{(1+\mu)(1-2\mu)} \begin{bmatrix} 1-\mu & \mu & \mu & 0 & 0 & 0 \\ & 1-\mu & \mu & 0 & 0 & 0 \\ & & 1-\mu & 0 & 0 & 0 \\ & & & \dfrac{1-2\mu}{2} & 0 & 0 \\ & & & & \dfrac{1-2\mu}{2} & 0 \\ & & & & & \dfrac{1-2\mu}{2} \end{bmatrix} = \begin{Bmatrix} \varepsilon_x \\ \varepsilon_y \\ \varepsilon_z \\ \gamma_{xy} \\ \gamma_{yz} \\ \gamma_{xz} \end{Bmatrix}$$

In case of two dimensional elasticity, the above relations get further simplified. There are two types of two dimensional elastic problems, namely plane stress and plane strain problems.

Stress

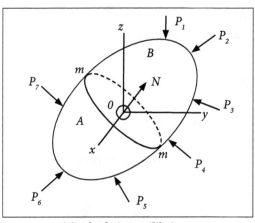

(a) A body in equilibrium.

The figure (a) represent a body in equilibrium. Under the action of external forces P_1, . . ., P_7 internal forces will be produced between the parts of the body.

To study the magnitude of these forces at any point O, let us imagine the body divided into two parts A and B by a cross section mm through this point.

Considering one of these parts, for instance, A, it can be stated that it is in equilibrium under the action of external forces P_1,...,P_7 and the inner forces distributed over the cross section mm and representing the actions of the material of the part B on the material of the part A. It will be assumed that these forces are continuously distributed over the area mm in the same way that hydrostatic pressure or wind pressure is continuously distributed over the surface on which it acts.

The magnitudes of such forces are usually defined by their intensity, i.e., by the amount of force per unit area of the surface on which they act. In discussing internal forces this intensity is called stress.

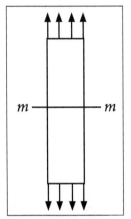

b) Prismatical bar subjected to uniformly distributed over the ends.

In the simplest case of a prismatical bar submitted to tension by forces uniformly distributed over the ends shown in (Figure b), the internal forces are also uniformly distributed over any cross section in mm. Hence the intensity of this distribution, i.e., the stress, can be obtained by dividing the total tensile force P by the cross-sectional area A.

In the case just considered the stress was uniformly distributed over the cross section. In the general case of the figure (a) the stress is not uniformly distributed over mm.

To obtain the magnitude of stress acting on a small area δA, cut out from the cross section mm at any point O, we assume that the forces acting across this elemental area, due to the action of material of the part B on the material of the part A can be reduced to a resultant δP.

If we now continuously contract the elemental area δA, the limiting value of the ratio $\delta P/\delta A$ gives us the magnitude of the stress acting on the cross section mm at the point O. The limiting direction of the resultant δP is the direction of the stress.

In the general case the direction of stress is inclined to the area δA on which it acts and we usually resolve it into two components, a normal stress perpendicular to the area and shearing stress acting in the plane of the area δA.

Notation for Forces and Stresses

There are two kinds of external forces which may act on bodies. Forces distributed over the surface of the body, such as the pressure of one body on another or hydrostatic pleasure, are called surface forces.

Forces distributed over the volume of a body, such as gravitational forces, magnetic forces or in the case of a body in motion, inertia forces, are called body forces.

The surface force per unit area we shall usually resolve into three components parallel to the coordinate axes and use for these components the notation $\overline{X}, \overline{Y}, \overline{Z}$. We shall also resolve the body force per unit volume into three components and denote these components by X, Y, Z.

We shall use the letter σ for denoting normal stress and the letter τ for shearing stress. To indicate the direction of the plane on which the stress is acting, subscripts to these letters are used.

If we take a very small cubic element at a point O (Figure a) with sides parallel to the coordinate axes, the notations for the components of stress acting on the sides of this element and the directions taken as positive are as indicated in the figure (C).

For the sides of the element perpendicular to the y-axis, for instance, the normal components of stress acting on these sides are denoted by σ_y. The subscript y indicates that the stress is acting on a plane normal to the y-axis. The normal stress is taken positive when it produces tension and negative when it produces compression.

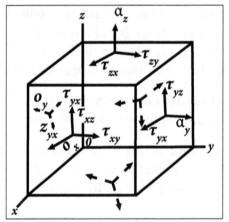

(c) Notations for the components of stress.

The shearing stress is resolved into two components parallel to the coordinate axes. Two subscript letters are used in this case, the first indicating the direction of the normal to the plane under consideration and the second indicating the direction of the component of the stress.

For instance, if we again consider the sides perpendicular to the y-axis, the component in the x-direction is denoted by τ_{yx} and that in the z direction by τ_{yz}. The positive directions of the components of shearing stress on any side of the cubic element are taken as the positive directions of the coordinate axes if a tensile stress on the same side would have the positive direction of the corresponding axis.

If the tensile stress has a direction opposite to the positive axis, the positive direction of the shearing-stress components should be reversed. Following this rule the positive directions of all the components of stress acting on the right side of the cubic element as shown in (Figure c) coincide with the positive directions of the coordinate axes. The positive directions are all reversed if we are considering the left side of this element.

Components of Stress

For each pair of parallel sides of a cubic element, such as in the figure (c), one symbol is needed to denote the normal component of stress and two more symbols to denote the two components of shearing stress.

To describe the stresses acting on the six sides of a cubic element three symbols, σ_x, σ_y, σ_z are necessary for normal stresses and six symbols $\tau_{xy}, \tau_{yx}, \tau_{xz}, \tau_{zx}, \tau_{yz}, \tau_{zy}$ for shearing stresses.

By a simple consideration of the equilibrium of the element the number of symbols for shearing stresses can be reduced to three.

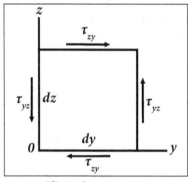

(d) Surface stresses.

If we take the moments of the forces acting on the element about the x-axis, for instance, only the surface stresses shown in the figure (d) need be considered.

Body forces, such as the weight of element, can be neglected in this instance, which follows from the fact that in reducing the dimensions of the element the body forces acting on it diminish as the cube of the linear dimensions while the surface forces diminish as the square of the linear dimensions.

Hence, for a very small element, body forces are small quantities of higher order than surface forces and can be neglected in calculating the surface forces. Similarly, moments due to non-uniformity of distribution of normal forces are of higher order than those due to the shearing forces and the limit.

Also, the forces on each side can be considered to be the area of the side times the stress at the middle. Then denoting the dimensions of the small element in the figure (d) by dx, dy, dz, the equation of equilibrium of this element, taking moments of forces about the x-axis, is given by:

$$\tau_{xy} \, dx \, dy \, dz = \tau_{yz} \, dx \, dy \, dz$$

The two other equations can be obtained in the same manner. From these equations we find,

$$\tau_{xy} = \tau_{yx}, \; \tau_{zx} = \tau_{xz}, \; \tau_{zy} = \tau_{yz} \qquad \qquad ...(1)$$

Hence, for two perpendicular sides of a cubic element the components of shearing stress perpendicular to the line of intersection of these sides are equal.

The six quantities σ_x, σ_y, σ_z, $\tau_{xy} = \tau_{yx}$, $\tau_{zx} = \tau_{xz}$, $\tau_{zy} = \tau_{yz}$ are therefore sufficient to describe the stresses acting on the coordinate planes through a point; these will be called the components of stress at the point.

Strain

Components of Strain

The deformation of an elastic body it will be assumed that there are enough constraints to prevent the body from moving as a rigid body, so that no displacements of particles of the body are possible without a deformation of it.

The small displacements of particles of a deformed body will usually be resolved into components u, v, w to parallel to the coordinate axes x, y, z, respectively. It will be assumed that these components are very small quantities varying continuously over the volume of the body.

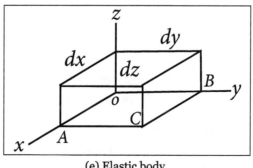

(e) Elastic body.

Consider a small element dx dy dz of an elastic body as shown in Figure (e). If the body undergoes a deformation and u, v, w are the components of the displacement of the point O, the displacement in the x-direction of an adjacent point A on the x-axis is due to the increase $(\partial u/\partial x)\, dx$ of the function u with increase of the coordinate x.

$$u + \frac{\partial u}{\partial x} dx$$

The increase in length of the element OA due to deformation is therefore $(\partial u/\partial x)\, dx$. Hence, the unit elongation at point O in the x-direction is $\partial u/\partial x$. In the same manner it can be shown that the unit elongations in the y- and z-directions are given by the derivatives $\partial v/\partial y$ and $\partial w/\partial z$.

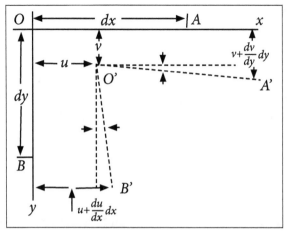

(f) The distortion of the angle between the elements OA and OB.

Let us consider now the distortion of the angle between the elements OA and OB, the figure (f). If u and v are the displacements of the point O in the x- and y-directions, the displacement of the point A in the y-direction and of the point B in the x-direction are v+ ($\partial v/\partial x$) dx and u +($\partial u/\partial y$) dy respectively.

Due to these displacements the new direction O'A' of the element OA is inclined to the initial direction by the small angle indicated in the figure, equal to $\partial v/\partial x$. In the same manner the direction O'B' is inclined to OB by the small angle $\partial u/\partial y$.

From this it will be seen that the initially right angle AOB between the two elements OA and OB is diminished by the angle $\partial v/\partial x + \partial u/\partial y$. This is the shearing strain between the planes xz and yz. The shearing strains between the planes xy and xz and the planes yx and yz can be obtained in the same manner.

We shall use the letter ϵ for unit elongation and the letter ϵy for unit shearing strain. To indicate the directions of strain we shall use the same subscripts to these letters as for the stress components.

$$\varepsilon_x = \frac{\partial u}{\partial x} \qquad \varepsilon_y = \frac{\partial v}{\partial y} \qquad \varepsilon_z = \frac{\partial w}{\partial z}$$

$$\gamma_{xy} = \frac{\partial u}{\partial y} + \frac{\partial v}{\partial x}, \quad \gamma_{xz} = \frac{\partial u}{\partial z} + \frac{\partial w}{\partial x} \quad \gamma_{yz} = \frac{\partial v}{\partial z} + \frac{\partial w}{\partial y} \qquad \dots(2)$$

The six quantities $\varepsilon_x, \dots, \gamma_{yz}$ are called the components of strain.

Constitutive Relations

Hooke's Law

The relations between the components of stress and the components of strain have been established experimentally and are known as Hooke's law.

Imagine an elemental rectangular parallelepiped with the sides parallel to the coordinate axes and submitted to the action of normal stress σ_z, uniformly distributed over two opposite sides.

Experiments show that in the case of an isotropic material these normal stresses do not produce any distortion of angles of the element. The magnitude of the unit elongation of the element is given by the equation (a) in which E is the modulus of elasticity in tension.

$$\varepsilon_x = \frac{\sigma_x}{E} \qquad \qquad ...(a)$$

Materials used in engineering structures have moduli which are very large in comparison with allowable stresses and the unit elongation (a) is a very small quantity. In the case of structural steel, for instance, it is usually smaller than 0.001.

Extension of the element in the x-direction is accompanied by lateral contractions:

$$\varepsilon_y = -v\frac{\sigma_x}{E}, \qquad \qquad \varepsilon_z = -v\frac{\sigma_x}{E} \qquad \qquad ...(b)$$

In which v, is a constant called Poisson's ratio.

For many materials Poisson's ratio can be taken equal to 0.25. For structural steel it is usually taken equal to 0.3. Equations (a) and (b) can be used also for simple compression. Within the elastic limit the modulus of elasticity and Poisson's ratio in compression are the same as in tension.

If the above element is submitted to the action of normal stresses $\sigma_x, \sigma_y, \sigma_z$ uniformly distributed over the sides, the resultant components of strain can be obtained by using equations (a) and (b).

Experiments show that to get these components we have to superpose the strain components produced by each of the three stresses. By this method of superposition we obtain the equations:

$$\varepsilon_x = \frac{1}{E}\left[\sigma_x - v\left(\sigma_y + \sigma_z\right)\right]$$

$$\varepsilon_y = \frac{1}{E}\left[\sigma_y - v\left(\sigma_y + \sigma_z\right)\right]$$

$$\varepsilon_z = \frac{1}{E}\left[\sigma_z - v\left(\sigma_x + \sigma_y\right)\right] \qquad \qquad ...(3)$$

This method of superposition used in calculating total deformations and stresses produced by several forces. This method is legitimate as long as the deformations are small and the corresponding small displacements do not affect substantially the action of the external forces.

In such cases we neglect small changes in dimensions of deformed bodies and also small displacements of the points of application of external forces and base our calculations on initial dimensions and initial shape of the body.

The resultant displacements will then be obtained by superposition in the form of linear functions of external forces, as in deriving the equations (3).

There are, however, exceptional cases in which small deformations cannot be neglected but must be taken into consideration. As an example of this kind the case of the simultaneous action on a thin bar of axial and lateral forces may be mentioned.

Axial forces alone produce simple tension or compression, but they may have a substantial effect on the bending of the bar if they are acting simultaneously with lateral forces.

In calculating the deformation of bars under such conditions, the effect of the deflection on the moment of the external forces must be considered, even though the deflections are very small. Then the total deflection is no longer a linear function of the forces and cannot be obtained by simple superposition.

Equations (3) show that the relations between elongations and stresses are completely defined by two physical constants E and υ. The same constants can also be used to define the relation between shearing strain and shearing stress.

Let us consider the particular case of deformation of the rectangular parallelopiped in which $\sigma_y = -\sigma_z$ and $\sigma_x = 0$. Cutting out an element ab c d by planes parallel to the x-axis and at 45° to the y- and z-axes Figure (g), by summing up the forces along and perpendicular to bc, that the normal stress on the sides of this element is zero and the shearing stress on the sides is given by:

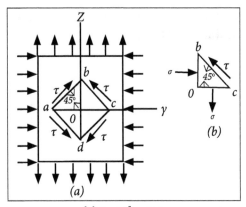

(g) y- and z-axes

$$\Upsilon = \frac{1}{2}\left(\sigma_z - \sigma_y\right) = \sigma_z \qquad \qquad ...(c)$$

Such a condition of stress is called pure shear. The elongation of the vertical element ob is equal to the shortening of the horizontal elements oa and oc and neglecting a small quantity of the second order we conclude that the lengths ab and bc of the element do not change during deformation.

The angle between the sides ab and bc changes and the corresponding magnitude of shearing strain γ may he found from the triangle obc. After deformation, we have:

$$\frac{oc}{ob}=\tan\left(\frac{\pi}{4}-\frac{\gamma}{2}\right)=\frac{1+\varepsilon_y}{1+\varepsilon_z}$$

Substituting, from the equations (3):

$$\varepsilon_z=\frac{1}{E}\left(\sigma_z-v\sigma_y\right)=\frac{(1+v)\sigma_z}{E}$$

$$\varepsilon_y=-\frac{(1+v)\sigma_z}{E}$$

And nothing that for small γ:

$$\tan\left(\frac{\pi}{4}-\frac{\gamma}{2}\right)=\frac{\tan\dfrac{\pi}{4}-\tan\dfrac{\gamma}{2}}{1+\tan\dfrac{\pi}{4}\tan\dfrac{\gamma}{2}}=\frac{1-\dfrac{\gamma}{2}}{1+\dfrac{\gamma}{2}}$$

We find:

$$\gamma=\frac{2(1+v)\sigma_z}{E}=\frac{2(1+v)\Upsilon}{E} \qquad\qquad ...(4)$$

Thus, the relation between shearing strain and shearing stress is defined by the constants E and υ. Often the notation is used,

$$G=\frac{E}{2(1+v)} \qquad ...\,5 \qquad\qquad ...(5)$$

Then, the equation (4) becomes:

$$\gamma=\frac{\tau}{G}$$

The constant G, defined by (5), is called the modulus of elasticity in shear or the modulus of rigidity.

If shearing stresses act on the sides of an element, as shown in the figure (c), the distortion of the angle between any two coordinate axes depends only on shearing-stress components parallel to these axes and we obtain,

$$\gamma_{xy} = \frac{1}{G}\tau_{xy}, \qquad \gamma_{yz} = \frac{1}{G}\tau_{yz}, \qquad \gamma_{zx} = \frac{1}{G}\tau_{zx} \qquad \ldots(6)$$

The elongations (3) and the distortions (6) are independent of each other. Hence the general case of strain, produced by three normal and three shearing components of stress, can be obtained by superposition: on the three elongations given by equations (3) are superposed three shearing strains given by equations (6).

Equations (3) and (6) give the components of strain as functions of the components of stress. Sometimes the components of stress expressed as functions of the components of strain are needed. These can be obtained as follows. Adding equations (3) together and using the notations,

$$e = \varepsilon_x + \varepsilon_y + \varepsilon_z$$

$$\theta = \sigma_x + \sigma_y + \sigma_z \qquad \ldots(7)$$

We obtain the following relation between the volume expansion e and the sum of normal stresses:

$$e = \frac{1-2v}{E}\theta \qquad \ldots(8)$$

In the case of a uniform hydrostatic pressure of the amount p we have:

$$\sigma_x = \sigma_y = \sigma_z = -p$$

And equation (8) gives:

$$e = -\frac{3(1-2v)p}{E}$$

Which represents the relation between unit volume expansion e and hydrostatic pressure p.

The quantity $E/3(1-2\upsilon)$ is called the modulus of volume expansion.

Using notations (7) and solving equations (3) for σ_x, σ_y, σ_z we find,

$$\sigma_x = \frac{vE}{(1+v)(1-2v)}e + \frac{E}{1+v}\epsilon_x$$

$$\sigma_y = \frac{vE}{(1+v)(1-2v)}e + \frac{E}{1+v}\epsilon_y$$

$$\sigma_z = \frac{v\,E}{(1+v)(1-2v)} e + \frac{E}{1+v} \epsilon_z \qquad \ldots(9)$$

or using the notation,

$$\lambda = \frac{v\,E}{(1+v)(1-2v)} \qquad \ldots(10)$$

And equation (5), these become,

$$\sigma_x = \lambda_e + 2G\varepsilon$$

$$\sigma y = \lambda_e + 2G_y$$

$$\sigma z = \lambda_e + 2G\varepsilon_z \qquad \ldots(11)$$

2.3   Plane Stress and Plane Strain Problems

Plane stress is defined to be a state of stress in which the normal stress and the shear stresses directed perpendicular to the plane are assumed to be zero. For instance, in figures (a) and (b), the plates in the x-y plane shown subjected to surface tractions T in the plane are under a state of plane stress that is, the normal stress σz and the shear stresses τ_{xy} and τ_{yz}, are assumed to be zero.

Generally, members that are thin (those with a small z dimension compared to the in-plane x and y dimensions) and whose loads act only in the x-y plane can be considered to be under plane stress.

Plane Strain

Plane strain is defined to be a state of strain in which the strain normal to the x-y plane ε_z and the shear strains γ_{xz} and γ_{yz}, are assumed to be zero. The assumptions of plane strain are realistic for long bodies with constant cross-sectional area subjected to loads that act only in the x- or y- directions and do not vary in the z- direction.

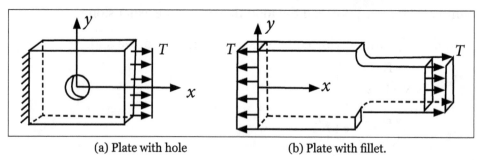

(a) Plate with hole   (b) Plate with fillet.

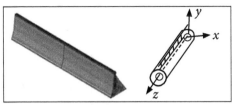

(a) Dam subjected to horizontal loading   (b) pipe subjected to a vertical load.

Some plane strain examples are shown in the figure above. In these examples, only a unit thickness (1 cm or 1 m) of the structure is considered because each unit thickness behaves identically. The finite element models of the structures shown in the figure above consist of appropriately discretized cross sections in the x-y plane with the loads acting over unit thicknesses in the x and/or y directions only.

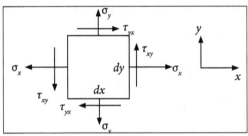

Two dimensional state of stress.

In a stressed body, the values of stresses change from face to face of an element. Hence, on positive face the various stresses act with superscript '+'.

Face	Stress on −ve Face	Stress on +ve Face
X	σ_x	$\sigma_x^+ = \sigma_x + \dfrac{\partial \sigma_x}{\partial_x} dx$
	τ_{xy}	$\tau_{xy}^+ = \tau_{xy} + \dfrac{\partial \tau_{xy}}{\partial_x} dx$
	τ_{yz}	$\tau_{xz}^+ = \tau_{xz} + \dfrac{\partial \tau_{xz}}{\partial_x} dx$
Y	σ_y	$\sigma_y^+ = \sigma_y + \dfrac{\partial \sigma_y}{\partial_y} dy$
	τ_{yx}	$\tau_{yx}^+ = \tau_{yx} + \dfrac{\partial \tau_{yx}}{\partial_y} dy$
	τ_{yz}	$\tau_{yz}^+ = \tau_{yz} + \dfrac{\partial \tau_{yz}}{\partial_y} dy$

	σ_z	$\sigma_z^+ = \sigma_z + \dfrac{\partial \sigma_z}{\partial_z} dz$
z	τ_{zx}	$\tau_{zx}^+ = \tau_{zx} + \dfrac{\partial \tau_{zx}}{\partial_z} dz$
	τ_{zy}	$\tau_{zy}^+ = \tau_{zy} + \dfrac{\partial \tau_{zy}}{\partial_z} dz$

Let the intensity of body forces acting on the element in x, y, z directions be X, Y and Z respectively. The intensity of body forces are uniform over entire body. Hence the total body force in x, y, z direction on the element shown is given by:

(i) $d_x d_y d_z$ in x – Direction.

(ii) $d_x d_y d_z$ in y – Direction.

(iii) $d_x d_y d_z$ in z – Direction.

Plane Stress Problems

The thin plates subject to forces in their plane only, fall under this category of the problems. In this, there is no force in the z-direction and no variation of any forces in z-direction. Hence,

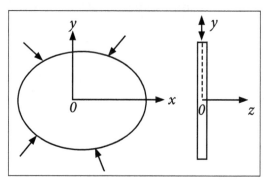

$$\sigma_z = \tau_{xz} = \tau_{yz} = 0$$

The condition $\tau_{xz} = \tau_{yz} = 0$ give $\gamma_{xz} = \gamma_{yz} = 0$ and the condition $\sigma_z = 0$ gives,

$$\sigma_z = \mu\varepsilon_x + \mu\varepsilon_y + (1-\mu)\varepsilon_z = 0$$

$$\varepsilon_z = -\frac{\mu}{1-\mu}\left(\varepsilon_z + \varepsilon_y\right)$$

$$\begin{Bmatrix} \sigma_x \\ \sigma_y \\ \sigma_z \end{Bmatrix} = \frac{E}{1-\mu^2} \begin{bmatrix} 1 & \mu & 0 \\ \mu & 1 & 0 \\ 0 & 0 & \dfrac{1-\mu}{2} \end{bmatrix} = \begin{Bmatrix} \varepsilon_x \\ \varepsilon_y \\ \gamma_{xy} \end{Bmatrix}$$

Plain Strain Problems

A long body subject to significant lateral forces falls under this category of problems. Examples of such problems are pipes, retaining walls, long strip footings, gravity dams, tunnels. In these problems, except for a small distance at the ends, state of stress is represented by any small longitudinal strip. The displacement in longitudinal direction (z-direction) is zero in typical strip.

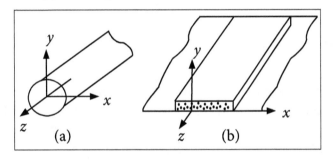

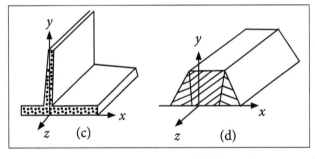

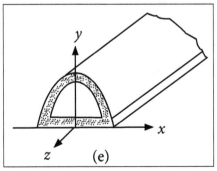

Hence, the strain components are given by:

$$\varepsilon_z = \gamma_{xz} = \gamma_{yz} = 0$$

$\gamma_{xz} = \gamma_{yz} = 0$ means τ_{xz} and τ_{yz} are zero

$\varepsilon_z = 0$ means

$$\varepsilon_z = \frac{\sigma_z}{E} - \mu \frac{(\sigma_x + \sigma_Y)}{E} = 0$$

$$\sigma_z = \mu(\sigma_x + \sigma_Y)$$

$$\begin{Bmatrix} \sigma_x \\ \sigma_y \\ \tau_{xy} \end{Bmatrix} = \frac{E}{(1+\mu)(1-2\mu)} \begin{bmatrix} 1-\mu & \mu & 0 \\ \mu & 1-\mu & 0 \\ 0 & 0 & \frac{1-2\mu}{2} \end{bmatrix} = \begin{Bmatrix} \varepsilon_x \\ \varepsilon_y \\ \gamma_{xy} \end{Bmatrix}$$

2.3.1 Different Methods of Structural Analysis Including Numerical Methods

Analysis of any structure may be performed based on some assumptions. These assumptions reflect the purpose and features of the structure, type of loads and operating conditions, properties of materials, etc. In whole, structural analysis may be divided into three large principal groups. They are static analysis, stability and vibration analysis.

Static analysis presumes that the loads act without any dynamical effects. Moving loads imply that only the position of the load is variable. Static analysis combines the analysis of a structure from a viewpoint of its strength and stiffness.

Static Linear Analysis (SLA)

The purpose of this analysis is to determine the internal forces and displacements due to time-independent loading conditions. This analysis is based on following conditions:

- Material of a structure obeys Hook's law.

- Displacements of a structure are small.

- All constraints are two-sided – It means that if constraint prevents displacement in some direction then this constraint prevents displacement in the opposite direction as well.

- Parameters of a structure do not change under loading.

Nonlinear Static Analysis

The purpose of this analysis is to determine the displacements and internal forces due to time-independent loading conditions, as if a structure is nonlinear. There are different types of nonlinearities. They are physical (material of a structure does not obey Hook's law), geometrical (displacements of a structure are large), structural (structure with gap or constraints are one-sided, etc.) and mixed nonlinearity.

Stability analysis deals with structures which are subjected to compressed time independent forces.

Buckling Analysis

The purpose of this analysis is to determine the critical load (or critical loads factor) and corresponding buckling mode shapes.

P-Delta Analysis

For tall and flexible structures, the transversal displacements may become significant. Therefore we should take into account the additional bending moments due by axial compressed loads P on the displacements caused by the lateral loads. In this case, we say that a structural analysis is performed on the basis of the deformed design diagram.

Dynamical Analysis

Dynamical analysis means that the structures are subjected to time-dependent loads, the shock and seismic loads, as well as moving loads with taking into account the dynamical effects.

Free-Vibration Analysis

The purpose of this analysis is to determine the natural frequencies (eigenvalues) and corresponding mode shapes (eigen functions) of vibration. This information is necessary for dynamical analysis of any structure subjected to arbitrary dynamic load, especially for seismic analysis. FVA may be considered for linear and nonlinear structures.

Stressed Free-Vibration Analysis

The purpose of this analysis is to determine the eigenvalues and corresponding eigen functions of a structure, which is subjected to additional axial time-independent forces.

Time-History Analysis

The purpose of this analysis is to determine the response of a structure, which is subjected to arbitrarily time-varying loads.

Fundamental Assumptions of Structural Analysis

Analysis of structures that is based on the following assumptions is called the elastic analysis:

- Material of the structure is continuous and absolutely elastic.

- Relationship between stress and strain is linear.

- Deformations of a structure, caused by applied loads, are small and do not change original design diagram.

- Superposition principle is applicable.

Superposition principle means that any factor, such as reaction, displacement, etc., caused by different loads which act simultaneously, are equal to the algebraic or geometrical sum of this factor due to each load separately.

For example, reaction of a movable support under any loads has one fixed direction. So the reaction of this support due to different loads equals to the algebraic sum of reactions due to action of each load separately.

Vector of total reaction for a pinned support in case of any loads has different directions, so the reaction of pinned support due to different loads equals to the geometrical sum of reactions, due to action of each load separately.

Fundamental Approaches of Structural Analysis

- There are two fundamental approaches to the analysis of any structure. The first approach is related to analysis of a structure subjected to given fixed loads and is called the fixed loads approach.

- The results of this analysis are diagrams, which show a distribution of internal forces (bending moment, shear and axial forces) and deflection for the entire structure due to the given fixed loads. These diagrams indicate the most unfavorable point (or member) of a structure under the given fixed loads.

- The second approach assumes that a structure is subjected to unit concentrated moving load only. This load is not a real one but imaginary. The results of the second approach are graphs called the influence lines.

- Influence lines are plotted for reactions, internal forces, etc. Internal forces diagrams and influence lines have a fundamental difference. Each influence line shows distribution of internal forces in the one specified section of a structure due to location of imaginary unit moving load only.

- These influence lines indicate the point of a structure where a load should be placed in order to reach a maximum (or minimum) value of the function under consideration at the specified section.

- It is very important that the influence lines may be also used for analysis of structure subjected to any fixed loads. Moreover, in many cases they turn out to be a very effective tool of analysis.

- Influence lines method presents the higher level of analysis of a structure, than the fixed load approach. Good knowledge of influence lines approaches an immeasurable increase in understanding of behavior of structure.

- Analyst, who combines both approaches for analysis of a structure in engineering practice, is capable to perform a complex analysis of its behavior. Both approaches do not exclude each other.

- In contrast, in practical analysis both approaches complement each other. Therefore, learning these approaches to the analysis of a structure will be provided in parallel way.

Numerical Methods

- While the derivation of the governing equations for most problems is not unduly difficult, their solution by exact methods of analysis is often difficult due to geometric and material complexities.

- In such cases, numerical methods of analysis provide alternative means of finding solutions. By a numerical simulation of a process, we mean the solution of the governing equations (or mathematical model) of the process using a numerical method and a computer.

- Numerical methods typically transform differential equations governing a continuum to a set of algebraic equations of a discrete model of the continuum that are to be solved using computers.

- There exist a number of numerical methods, many of which are developed to solve differential equations. In the finite difference approximation of a differential equation, the derivatives in the latter are replaced by difference quotients that involve the values of the solution at discrete mesh points of the domain.

- The resulting algebraic equations are solved for the values of the solution at the mesh points after imposing the boundary conditions. These ideas are illustrated with the help of two examples, one for an initial-value problem and another for a boundary-value problem.

- In the solution of a differential equation by a classical variational method, the equation is put into an equivalent weighted-integral form and then the approximate solution over the domain is assumed to be a linear combination $\left(\sum_j c_j \phi_j \right)$ of appropriately chosen approximation functions ϕ_j and undetermined coefficients c_j.

- The coefficients c_j are determined such that the integral statement equivalent to the original differential equation is satisfied. Various variational methods, e.g., the Ritz. Galerkin.

- Collocation and least-squares methods differ from each other in the choice of the integral form, weight functions and/or approximation functions.

- The classical variational methods, which are truly mesh-less methods are powerful methods that provide globally continuous solutions but suffer from the disadvantage that the approximation functions for problems with arbitrary domains are difficult to construct (the modern mesh-less methods seem to provide a way to construct approximation functions for arbitrary domains).

2.4 Basics of Finite Element Method

In the finite element method, the exact continuum or the body of matter, such as a solid, liquid or gas, is represented as an assemblage of subdivisions known finite elements. These elements are considered to be interconnected at the specified joints called as nodes or nodal points. The nodes generally lie on the element boundaries where the adjacent elements are considered to be connected.

Since, the actual variation of the field variable (e.g., stress, displacement, temperature, velocity or pressure) inside the continuum is not known, we assume that the variation of the field variable inside a finite element may be approximated by a simple function. These approximating functions are defined in terms of the values of the field variables at the nodes.

When the field equations for the whole continuum are written, the new unknowns will be the nodal values of the field variable. By solving the field equations, which are usually in the form of matrix equations, the nodal values of the field variable will be known. Once these are known, the approximating functions define the field variable throughout the assemblage of elements.

The solution of the general continuum problem by the finite element method always follows an orderly step-by-step process. With reference to static structural problems, the step-by-step procedure can be stated as follows.

Finite Element Analysis Involves the Following Steps:

- Discretization: It is the process of dividing the domain, structure or continuum into sub regions or subdivisions called finite elements. Element having a simple shape with which a domain or body or structure is discretized.

- Nodes are defined for each element and nodes are the locations or discrete points at which the unknown variables are to be determined. These unknown variables are called field variables as the unknown variable may be displacement, temperature or velocity depending on the type of problem under consideration. Collection of elements is called mesh. Elements are connected at the nodes.

- Approximation of field variable in an element: Polynomial expressions are normally employed to define the variation of the field variable. It is easy to differentiate and integrate the terms in polynomial expressions.

• Formulation of element equations: Element equations will be derived by minimization of a functional. The functional is an integral expression and minimization with respect to nodal variables yields the element equations.

The formulation of element equations can be accomplished by using one of the following methods:

• Variation formulation using calculus of variation.

• Weighted residual methods out of which Galerkin's method is extensively used.

For structural mechanics problems, the potential energy expression written in an integral form provides the necessary functional. Minimization of potential energy expression with respect to nodal displacements will yield element equations which are known as equations of equilibrium.

• Principle of minimum potential energy: Among all admissible configurations of a conservative system, those that satisfy the equations of equilibrium make the potential energy stationary with respect to small admissible variations of displacement.

• Assembly of matrices to form global or system equations: The stiffness matrix and load vector (also called force vector) for each element are formulated. Assembling these matrices for all elements will give global stiffness matrix and global load vector to formulate the global or system equations.

$$\left[K\{e\}\right]\{q(e)\}=\{F(e)\}$$

$$\sum_{c=1}^{n}\left(\left[K^{(e)}\right]\{q^{(e)}\}=\{F^{(e)}\}\right)=0$$

$$[K]\{Q\}\ =\ \{F\}$$

Where,

N-Number of elements.

$\left[\ ^{(\ )}\right]$-Element stiffness matrix.

$[K]$-Global stiffness matrix.

$\{Q\}$-Global nodal displacement vector.

$\{F^{(e)}\}$-Element load vector.

$\{q^{(e)}\}$-Element nodal displacement vector.

{F}-Global load vector.

- Solving the global or system equations will provide the solution for field variable i.e., displacements for a solid or structural mechanics problem.

$$\{Q\} = [K]^{-1}\{F\}$$

- Secondary quantities like strain and stress can be calculated once the nodal displacements are known.

The description of finite element method given above is based on displacement formulation for structural mechanics problem.

In addition to displacement formulation, other methods employed are:

- Equilibrium method
- Mixed method.

In equilibrium method, stress is field variable and displacement is to be derived from stress. When the formulation is based on stress, compatibility equations need to satisfy.

In mixed method, part of the domain is solved using displacement formulation and remaining part with equilibrium method. At the outset finite element analysis relies on two important functions.

- A continuous piecewise smooth function is needed to prescribe the field variable within the element.
- An integral expression called functional is used to generate element equations.

Even though the finite element method was originally used for structural mechanics problems, presently it is used in many areas including electrostatic, magnetic field problems as well as problems encountered in Biomedical engineering.

Since finite element method (FEM) has versatility to provide solutions pertaining to several fields. Application of FEM is well known in the field of automobile, aerospace, thermal, heat transfer and fluid flow areas.

Advantages of Finite Element Method:

- FEM can handle irregular geometry in a convenient manner.
- Higher order elements may be implemented.
- Non - homogeneous materials can be handled easily.
- Handles general load conditions without difficulty.

Disadvantages of Finite Element Method:

- It requires longer execution time compared with FEM.

- Output result will vary considerably.

- It requires a digital computer and fairly extensive.

Applications of Finite Element Analysis

Structural Problems:

- Stress concentration problems typically associated with holes, fillets or other changes in geometry in a body.

- Vibration Analysis: Example of vibration analysis is a beam that is subjected to different types of loading.

- Buckling Analysis: Example of buckling analysis is connecting rod which is subjected to axial compression.

- Stress analysis including truss and frame analysis.

Non-Structural Problems:

- Fluid flow analysis: Example of fluid flow analysis is fluid flow through pipes.

- Heat Transfer analysis: Example of heat transfer analysis is steady state thermal analysis on composite cylinder.

Geotechnical engineering: FEM applications include stress analysis, slope stability analysis, soil structure interactions, seepage of fluids in soils and rocks, analysis of dams, tunnels, bore holes, propagation of stress waves and dynamic soil structure interaction.

Fluid mechanics, hydraulic and water resources engineering: FEM applications include solutions of potential and viscous flow of fluids, steady and transient seepage in aquifers and porous media, movement of fluids in containers, external and internal flow analysis, seiche of lakes, ocean and harbours, salinity and pollution studies in surface and sub-surface water problems, sediment transport analysis and water distribution networks.

Mechanical engineering: In mechanical engineering, FEM applications include steady and transient thermal analysis in solids and fluids, stress analysis in solids, automotive design and analysis and manufacturing process simulation.

Nuclear engineering: FEM applications include steady and dynamic analysis of reactor containment structures, thermo-viscoelastic analysis of reactor components, steady and transient temperature-distribution analysis of reactors and related structures.

Electrical and electronics engineering: FEM applications include electrical network analysis, electro magnetics, insulation design analysis in high-voltage equipment's, thermo-sonic wire bond analysis, dynamic analysis of motors, moulding process analysis in encapsulation of integrated circuits and heat analysis in electrical and electronic equipment.

Metallurgical, chemical and environmental engineering: In metallurgical engineering, FEM is used for the metallurgical process simulation, moulding and casting. In chemical engineering, FEM can be used in the simulation of chemical processes, transport processes (including advection and diffusion) and chemical reaction simulations. FEM is used in environmental engineering widely in the areas of surface and sub-surface pollutant transport modeling, air pollution modeling, land-fill analysis and environmental process simulation.

Meteorology and bioengineering: In the recent times, FEM is used in climate predictions, monsoon prediction and wind predictions. FEM is also used in bioengineering for the simulation of various human organs, blood circulation prediction and even total synthesis of human body.

2.5   Different Steps involved in FEM

For simplicity's sake, for the presentation of the steps to follow, we will consider only the structural problem. Typically, for the structural stress-analysis problem, the engineer seeks to determine displacements and stresses throughout the structure, which is in equilibrium and is subjected to applied loads. For many structures, it is difficult to determine the distribution of deformation using conventional methods and thus the finite element method is necessarily used.

The analyst must make decisions regarding dividing the structure or continuum into finite elements and selecting the element type or types to be used in the analysis (step 1), the kinds of loads to be applied and the types of boundary conditions or supports to be applied. The other steps, 2–7, are carried out automatically by a computer program.

Step 1: Discretized and Select the Element Types

It involves dividing the body into an equivalent system of finite elements with associated nodes and choosing the most appropriate element type to model most closely the actual physical behavior.

The total number of elements used and their variation in size and type within a given body are primarily matters of engineering judgment.

The elements must be made small enough to give usable results and yet large enough to reduce computational effort. Small elements (and possibly higher-order elements) are generally desirable where the results are changing rapidly, such as where changes in geometry occur large elements can be used where results are relatively constant.

The discretized body or mesh is often created with mesh-generation programs or pre-processor programs available to the user.

The choice of elements used in a finite element analysis depends on the physical make-up of the body under actual loading conditions and on how close to the actual behavior the analyst wants the results to be.

Judgment concerning the appropriateness of one, two or three-dimensional ideal-izations is necessary. Moreover, the choice of the most appropriate element for a particular problem is one of the major tasks that must be carried out by the designer/analyst.

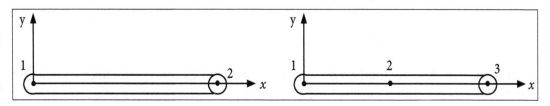

(a) Simple two-nodded line element (typically used to represent a bar or beam element) and the higher-order line element.

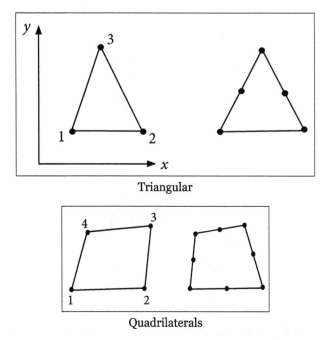

Triangular

Quadrilaterals

(b) Simple two-dimensional elements with corner nodes (typically used to represent plane stress/strain) and higher-order two-dimensional elements with intermediate nodes along the sides.

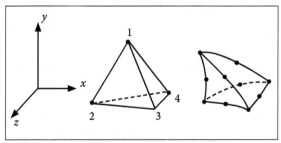

Tetrahedral

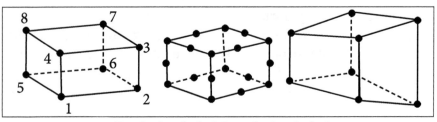

Regular hexahedral, Irregular hexahedral.

(c) Simple three-dimensional elements (typically used to represent three-dimensional stress state) and higher-order three-dimensional elements with intermediate nodes along edges.

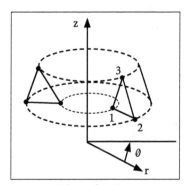

Triangular ring.

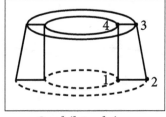

Quadrilateral ring.

(d) Simple axisymmetric triangular and quadrilateral elements used for axisymmetric problems.

(1) Various types of simple lowest-order finite elements with corner nodes only and higher-order elements with intermediate nodes.

Elements that are commonly employed in practice most of which are considered in here are shown in Figure. The primary line elements [Figure (a)] consist of bar (or truss) and beam elements. They have a cross-sectional area but are usually represented by line segments.

In general, the cross-sectional area within the element can vary, but throughout here it will be considered to be constant. These elements are often used to model trusses and frame structures.

The simplest line element (called a linear element) has two nodes, one at each end, although higher-order elements having three nodes [Figure (a)] or more (called quadratic, cubic, etc. elements) also exist.

The basic two-dimensional (or plane) elements [Figure (b)] are loaded by forces in their own plane (plane stress or plane strain conditions). They are triangular or quadrilateral elements.

The simplest two-dimensional elements have corner nodes only (linear elements) with straight sides or boundaries, although there are also higher-order elements, typically with midsize nodes [Figure (b)] (called quadratic elements) and curved sides.

The elements can have variable thicknesses throughout or be constant. They are often used to model a wide range of engineering problems. The most common three-dimensional elements [Figure (c)] are tetrahedral and hexahedral (or brick) elements, they are used when it becomes necessary to perform a three-dimensional stress analysis.

The basic three-dimensional elements have corner nodes only and straight sides, whereas higher-order elements with mid-edge nodes (and possible mid-face nodes) have curved surfaces for their sides [Figure (c)].

The axisymmetric element [Figure (d)] is developed by rotating a triangle or quadrilateral about a fixed axis located in the plane of the element through 360°. This element can be used when the geometry and loading of the problem are axisymmetric.

Step 2: Select a Displacement Function

It involves choosing a displacement function within each element. The function is defined within the element using the nodal values of the element. Linear, quadratic and cubic polynomials are frequently used functions because they are simple to work with in finite element formulation. However, trigonometric series can also be used.

For a two-dimensional element, the displacement function is a function of the co-ordinates in its plane (say, the x-y plane). The functions are expressed in terms of the nodal unknowns (in the two-dimensional problem, in terms of an x and a y component).

The same general displacement function can be used repeatedly for each element. Hence the finite element method is one in which a continuous quantity, such as the displacement throughout the body, is approximated by a discrete model composed of a set of piecewise-continuous functions defined within each finite domain or finite element.

Step 3: Define the Strain/Displacement and Stress/Strain Relationships

Strain/displacement and stress/strain relationships are necessary for deriving the equations for each finite element. In the case of one-dimensional deformation, say, in the x direction, we have strain εx related to displacement u by:

$$\varepsilon_x = \frac{du}{dx} \qquad \qquad ...(1)$$

For small strains. In addition, the stresses must be related to the strains through the stress/strain law-generally called the constitutive law. The ability to define the material behavior accurately is most important in obtaining acceptable results. The simplest of stress/strain laws, Hooke's law, which is often used in stress analysis, is given by:

$$\sigma_x = E\varepsilon_x \qquad \qquad ...(2)$$

Where,

σ_x -Stress in the x direction,

E-Modulus of elasticity.

Step 4: Derive the Element Stiffness Matrix and Equations

Initially, the development of element stiffness matrices and element equations was based on the concept of stiffness influence coefficients, which presupposes a background in structural analysis. We now present alternative methods used here that do not require this special background.

Direct Equilibrium Method:

According to this method, the stiffness matrix and element equations relating nodal forces to nodal displacements are obtained using force equilibrium conditions for a basic element, along with force/deformation relationships.

Because this method is most easily adaptable to line or one-dimensional elements and also this method used for spring, bar and beam elements.

Work or Energy Methods:

To develop the stiffness matrix and equations for two and three-dimensional elements, it is much easier to apply a work or energy method. The principle of virtual work (using virtual displacements), the principle of minimum potential energy and Castigliano's theorem are methods frequently used for the purpose of derivation of element equations.

The principle of virtual work is applicable for any material behavior, whereas the principle of minimum potential energy and Castigliano's theorem are applicable only to elastic materials.

Furthermore, the principle of virtual work can be used even when a potential function does not exist. However, all three principles yield identical element equations for linear-elastic materials, thus which method to use for this kind of material in structural analysis is largely a matter of convenience and personal preference.

For the purpose of extending the finite element method outside the structural stress analysis field, a functional' (a function of another function or a function that takes functions as its argument) analogous to the one to be used with the principle of minimum potential energy is quite useful in deriving the element stiffness matrix and equations.

For instance, letting π denote the functional and $f(x,y)$ denote a function f of two variables x and y, we then have $\pi = \pi f(x,y)$, where π is a function of the function f.

A more general form of a functional depending on two independent variables $u(x,y)$ and $v(x,y)$, where independent variables are x and y in Cartesian coordinates is given by:

$$\pi = \iint F\left(x,\ y,\ z,\ u,\ v,\ u_x, u_y, v_x, v_y, u_{xx}, \ldots, v_{yy}\right) dx\, dy \qquad \ldots(3)$$

Methods of Weighted Residuals:

The methods of weighted residuals are useful for developing the element equations; particularly popular is Galerkin's method. These methods yield the same results as the energy methods wherever the energy methods are applicable.

They are especially useful when a functional such as potential energy is not readily available. The weighted residual methods allow the finite element method to be applied directly to any differential equation.

Galerkin's method, along with the collocation, the least squares and the sub do-main weighted residual methods are obtained. To illustrate each method, they will all be used to solve a one-dimensional bar problem for which a known exact solution exists for comparison.

As the more easily adapted residual method, Galerkin's method will also be used to derive the bar element equations and the beam element equations and to solve the combined heat-conduction/convection/mass transport problem.

Using any of the methods just outlined will produce the equations to describe the behavior of an element. These equations are written conveniently in matrix form as:

$$\begin{Bmatrix} f_i \\ f_2 \\ f_3 \\ \vdots \\ f_n \end{Bmatrix} = \begin{bmatrix} k_{11} & k_{12} & k_{13} & \cdots & k_{1n} \\ k_{21} & k_{22} & k_{23} & \cdots & k_{2n} \\ k_{31} & k_{32} & k_{33} & \cdots & k_{3n} \\ \vdots & \vdots & & & \vdots \\ k_{n1} & & & \cdots & k_{nn} \end{bmatrix} \begin{Bmatrix} d_1 \\ d_2 \\ d_3 \\ \vdots \\ d_n \end{Bmatrix} \qquad \ldots (4)$$

or in compact matrix form as:

$$\{f\} = [k]\{d\} \qquad \ldots (5)$$

Where,

$\{f\}$ - The vector of element nodal forces,

$[k]$ -The element stiffness matrix (normally square and symmetric), and

$\{d\}$-The vector of unknown element nodal degrees of freedom or generalized displacements n.

Here generalized displacements may include such quantities as actual displacements, slopes or even curvatures.

Step 5: Assemble the Element Equations to Obtain the Global or Total Equations and Introduce Boundary Conditions

In this step the individual element nodal equilibrium equations generated in step 4 are assembled into the global nodal equilibrium equations. Direct method of superposition (called the direct stiffness method), whose basis is nodal force equilibrium, can be used to obtain the global equations for the whole structure.

Implicit in the direct stiffness method is the concept of continuity or compatibility, which requires that the structure remain together and that no tears occur anywhere within the structure.

The final assembled or global equation written in matrix form is:

$$\{F\} = [k]\{d\} \qquad \ldots (6)$$

Where,

$\{f\}$-The vector of global nodal forces,

[k]-The structure global or total stiff-ness matrix, (for most problems, the global stiffness matrix is square and symmetric), and

{d}-Now the vector of known and unknown structure nodal degrees of freedom or generalized displacements.

It can be shown that at this stage, the global stiffness matrix [k] is a singular matrix because its determinant is equal to zero. To remove this singularity problem, we must invoke certain boundary conditions (or constraints or supports) so that the structure remains in place instead of moving as a rigid body.

At this time it is sufficient to note that invoking boundary or support conditions results in a modification of the global equation (6). We also emphasize that the applied known loads have been accounted for in the global force matrix $\{f\}$.

Step 6: Solve for the Unknown Degrees of Freedom (or Generalized Displacements)

Equation (6), modified to account for the boundary conditions, is a set of simultaneous algebraic equations that can be written in expanded matrix form as:

$$
\begin{Bmatrix} F_1 \\ F_2 \\ \vdots \\ F_n \end{Bmatrix} = \begin{bmatrix} K_{11} & K_{12} & \cdots & K_{1n} \\ K_{21} & K_{22} & \cdots & K_{2n} \\ \vdots & & & \vdots \\ K_{n1} & K_{n2} & \cdots & K_{nn} \end{bmatrix} \begin{Bmatrix} d_1 \\ d_2 \\ \vdots \\ d_n \end{Bmatrix} \qquad ...(7)
$$

Where now n is the structure total number of unknown nodal degrees of freedom. These equations can be solved for the ds by using an elimination method (such as Gauss's method) or an iterative method (such as the Gauss–Seidel method).

These are called the primary unknowns; because they are the first quantities determined using the stiffness (or displacement) finite element method.

Step 7: Solve for the Element Strains and Stresses

For the structural stress-analysis problem, important secondary quantities of strain and stress (or moment and shear force) can be obtained because they can be directly expressed in terms of the displacements determined in step 6.

Typical relationships between strain and displacement and between stress and strain such as equations (1) and (2) for one-dimensional stress given in step 3 can be used.

Step 8: Interpret the Results

The final goal is to interpret and analyze the results for use in the design/analysis process. Determination of locations in the structure where large deformations and large stresses occur is generally important in making design/analysis decisions. Post processor computer programs help the user to interpret the results by displaying them in graphical form.

2.6   Different Approaches of FEM

There are two general direct approaches traditionally associated with the finite element method as applied to structural mechanics problems. One approach, called the force or flexibility method, uses internal forces as the unknowns of the problem.

To obtain the governing equations, first the equilibrium equations are used. Then necessary additional equations are found by introducing compatibility equations. The result is a set of algebraic equations for determining the redundant or unknown forces.

The second approach, called the displacement or stiffness method, assumes the displacements of the nodes as the unknowns of the problem. For instance, compatibility conditions requiring that elements connected at a common node, along a common edge or on a common surface before loading remain connected at that node, edge or surface after deformation takes place are initially satisfied. Then the governing equations are expressed in terms of nodal displacements using the equations of equilibrium and an applicable law relating forces to displacements.

These two direct approaches result in different unknowns (forces or displacements) in the analysis and different matrices associated with their formulations (flexibilities or stiff nesses). It has been shown that, for computational purposes, the displacement (or stiffness) method is more desirable because its formulation is simpler for most structural analysis problems.

Furthermore, a vast majority of general-purpose finite element programs have incorporated the displacement formulation for solving structural problems.

Another general method that can be used to develop the governing equations for both structural and nonstructural problems is the variational method. The variational method includes a number of principles.

One of these principles, because it is relatively easy to comprehend and is often introduced in basic mechanics courses, is the theorem of minimum potential energy that applies to materials behaving in a linear-elastic manner.

Another variational principle often used to derive the governing equations is the principle of virtual work. This principle applies more generally to materials that behave in a linear-elastic fashion, as well as those that behave in a nonlinear fashion.

The finite element method involves modeling the structure using small interconnected elements called finite elements. A displacement function is associated with each finite element. Every interconnected element is linked, directly or indirectly, to every other element through common (or shared) interfaces, including nodes and/or bound-ary lines and/or surfaces.

By using known stress/strain properties for the material making up the structure, one can determine the behavior of a given node in terms of the properties of every other element in the structure. The total set of equations describing the behavior of each node results in a series of algebraic equations best expressed in matrix notation.

2.6.1 Direct Approach and Energy Approach

There are number of ways in which one can formulate the properties of individual elements of the domain.

Most commonly used approaches to formulate element matrices are:

Direct Approach

The basic idea of finite element method was conceived from the physical procedure used in framed structural analysis and network analysis (pipe network and electric network).

It may be possible to choose elements in a way that leads to an exact representation of the problem in certain applications. Hence, element properties are derived from the fundamental physics and nature of the problem in the direct approach.

Analysis based on stiffness method is an example of this approach in structural mechanics. Main advantage of this approach is that an easy understanding of techniques and essential concepts is gained without much mathematical illustrations.

In solving a problem using the direct approach, first, the elements are defined and then their properties are determined. Once the elements have been selected, direct physical relationships are used to establish element equations in terms of concerned variables.

Finally, element equations for various elements or members are combined to generate a system of equations which are solved for the unknowns.

Some engineering problems which can be solved using the direct approach include spring systems, trusses, beams, fluid flow in pipe networks, electric resistance networks etc. Here, the linear spring system problem is analyzed to demonstrate the development of finite element technique using the direct approach formulation.

Linear Spring System

A system of linear springs is used in many engineering problems. To demonstrate the finite element procedure using the direct approach, a system of three linear springs as shown in the figure (1) is analyzed.

The system is connected in series with springs of different stiffness coefficients. One end of the spring on left-hand side (LHS) is rigidly fixed, while the other end is free to move, as shown in the figure (1a).

All springs can be subjected to either tension or compression. Physical parameters involved in this system are forces, displacements and spring stiff nesses.

In the finite element procedure, each spring is considered as an element and the connection as nodes. The system considered here has three springs and hence consists of three elements and four nodes as shown in the figure (1b).

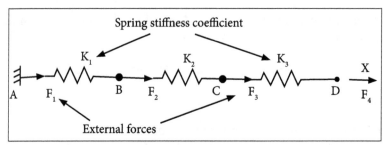

1(a) A linear spring system with three springs.

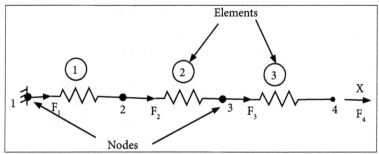

1(b) Finite element discretization of linear spring system.

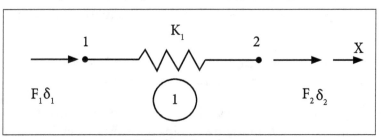

1(c) Single spring element.

Considering only one spring as shown in the figure (1c), which is being treated as isolated from system, force displacement equations can be developed as follows. Forces and displacements are defined at each node and field variable is the displacement.

There is no need for an interpolation function to represent variation of field variable over the element because an exact representation is available from Hooke's law, which relates the nodal displacements and the applied nodal forces F given as:

$$\delta = \frac{1}{k}F = CF \qquad \qquad ...(1)$$

Where,

k-The spring stiffness constant.

Δ-The displacement.

F-The force applied.

C-The deflection coefficient.

In other words, k can be treated as the force required producing unit deflection or C as the deflection caused by a unit force.

Force displacement equations for node 1 and node 2 are:

$$F_1 = k\delta_1 - k\delta_2 \qquad\qquad ...(2)$$

$$F_2 = -k\delta_1 + k\delta_2 \qquad\qquad ...(3)$$

In matrix form, equations (2) and (3) can be written as,

$$\begin{bmatrix} k & -k \\ -k & k \end{bmatrix} \begin{Bmatrix} \delta_1 \\ \delta_2 \end{Bmatrix} = \begin{Bmatrix} F_1 \\ F_2 \end{Bmatrix} \qquad\qquad ...(4)$$

$$\text{Or } \begin{bmatrix} k^e \end{bmatrix} \{q^e\} = \{f^e\} \qquad\qquad ...(5)$$

The square matrix $\begin{bmatrix} k^e \end{bmatrix}$ is known as the element stiffness matrix, the column vector $\{q^e\}$ is the nodal displacement vector, $\{f^e\}$ is the resultant nodal force vector for the element.

Depending on the problem, the stiffness coefficients of matrix $\begin{bmatrix} k^e \end{bmatrix}$ are determined exactly or approximately. In the present case, a typical stiffness coefficient k_{ij} of $\begin{bmatrix} k^e \end{bmatrix}$ is defined as the force required at node i to produce a unit deflection at node j.

It should be noted that the left hand side element is constrained to have zero displacement at one node which does not influence derivation of element matrix. Constraint conditions or boundary conditions are taken into account only after element equations are assembled to form system of equations.

Energy Approach

Principle of virtual work, the principle of minimum potential energy and Castigliano's theorem are used frequently to derive equations of elements used in stress analysis problems.

Here, the principle of minimum potential energy, which is applicable only to conservative system. Principle of minimum energy is based on the idea of finding the consistent states (equilibrium states) of the body or structure associated with stationary values of a scalar quantity assumed by the loaded bodies.

This scalar quantity is generally referred to as a 'functional' (defined to be a function of another function). For example, let $f(x)$ be a function of the variable x and π be the functional defined such that $\pi = \pi[f(x)]$. Then, π is a function of the function f.

Stationary values are given by (see figure 2):

$$\frac{d\pi}{dx} = 0 \qquad \qquad ...(6)$$

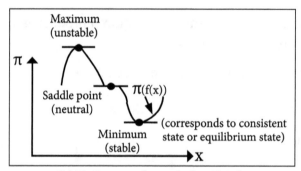

(2) Stationary values of a functional.

In structural analysis problems, π is the potential energy of the body or structure. In this case, element equations can be obtained by invoking the principle of minimum potential energy (minimizing the potential function).

FEM can be applied generally to any problem provided this scalar function is available. As an illustration, consider one spring system mentioned above. The potential energy for this elastic system is:

$$\pi = \frac{1}{2}k\left[\pm(\delta_1 - \delta_2)\right]^2 = \frac{1}{2}k\left(\delta_1^2 - 2\delta_1\delta_2 + \delta_2^2\right)$$

$$\frac{\partial\pi}{\partial\delta_1} = k\delta_1 - k\delta_2$$

$$\frac{\partial\pi}{\partial\delta_2} = -k\delta_1 + k\delta_2$$

The stiffness matrix can be written as before.

3

Element Properties

3.1 Interpolation Functions for General Element Formulations

The requirements for interpolation functions in terms of solution accuracy and convergence of a finite element analysis to the exact solution of a general field problem. Interpolation functions for various common element shapes in one, two and three dimensions are developed and these functions are used to formulate finite element equations for various types of physical problems.

The exception of the beam element, all the interpolation functions are applicable to finite elements used to obtain solutions to problems that are said to be Co-continuous. This terminology means that, across element boundaries. Only the zeroth-order derivatives of the field variable (i.e., the field variable itself) are continuous.

On the other hand, the beam element formulation is such that the element exhibits C1-continuity, since the first derivative of the transverse displacement (i.e., slope) is continuous across element boundaries. In general, in a problem having Cn-continuity, derivatives of the field variable up to and including nth-order derivatives are continuous across element boundaries.

Compatibility and Completeness Requirement

The line elements (spring, truss, and beam) illustrate the general procedures used to formulate and solve a finite element problem and are quite useful in analyzing truss and frame structures. Such structures, however, tend to be well defined in terms of the number and type of elements used. In most engineering problems, the domain of interest is a continuous solid body, often of irregular shape, in which the behavior of one or more field variables is governed by one or more partial differential equations.

The objective of the finite element method is to Discretization the domain into a number of finite elements for which the governing equations are algebraic equations. Solution of the resulting system of algebraic equations then gives an approximate solution to the problem.

In finite element analysis, solution accuracy is judged in terms of convergence as the element "mesh" is refined. There are two major methods of mesh refinement. In the first, known as h-refinement, mesh refinement refers to the process of increasing the number of elements used to model a given domain, consequently, reducing individual element size.

In the second method, p-refinement, element size is unchanged but the order of the polynomials used as interpolation functions is increased.

The objective of mesh refinement in either method is to obtain sequential solutions that exhibit asymptotic convergence to values representing the exact solution. The mathematical proofs of convergence of finite element solutions to correct solutions are based on a specific, regular mesh refinement procedure defined as:

- Although the proofs are based on regular meshes of elements, irregular or un-structured meshes (such as in the figure) can give very good results. In fact, use of unstructured meshes is more often the case, since the geometries being modeled us most often irregular.

- The auto-meshing features of most finite element software packages produce irregular meshes. An example illustrating regular h-refinement as well as solution convergence is shown in the figure (a), which depicts a rectangular elastic plate of uniform thickness fixed on one edge and subjected to a concentrated load on one corner.

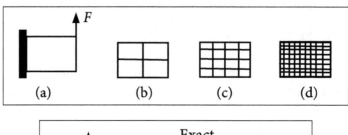

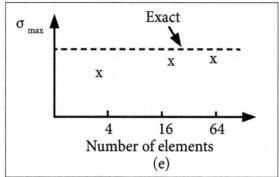

Example in convergence as elements mesh in refined.

This problem is modeled using rectangular plane stress elements and three meshes used in sequence, as shown in (Figure b, d). Solution convergence is depicted in the figure (e) in terms of maximum normal stress in the x direction. For example, the exact solution is taken to be the maximum bending stress computed using elementary beam theory.

The true exact solution is the plane stress solution from the theory of elasticity. However, the maximum normal stress is not appreciably changed in the elasticity solution.

The need for convergence during regular mesh refinement is rather clear. If convergence is not obtained, the engineer using the finite element method has absolutely no indication whether the results are indicative of a meaningful approximation to the correct solution. For a general field problem in which the field variable of interest is expressed on an element basis in the discretized form,

$$\phi^{(e)}(x,y,z) = \sum_{i=1}^{M} N_i(x,y,z)\phi_i \qquad \qquad ...(1)$$

Where, M -Number of element degrees of freedom:

$$AE\int_0^L \frac{dN_i}{dx}\frac{du}{dx}dx = \left[N_i\, AE\, \frac{du}{dx}\right]_0^L \qquad \qquad ...(2)$$

Integrating again by parts and rearranging gives:

$$EI_z \int_{x_1}^{x_2} \frac{d^2 N_i}{dx^2}\frac{d^2 v}{dx^2}dx = \int_{x_1}^{x_2} N_{iq}(x)dx - N_i\, EI_z\, \frac{d^3 v}{dx^3}$$

$$\left.\right|_{x_1}^{x_2} + \frac{dN_i}{dx}EI_z\frac{d^2 v}{dx^2}\Bigg|_{x_1}^{x_2}\quad i=1,4 \qquad \qquad ...(3)$$

The interpolation functions must satisfy two primary conditions to ensure convergence during mesh refinement the compatibility and completeness requirements.

3.1.1 Compatibility and Completeness

Compatibility

Along element boundaries, the field variable and its partial derivatives up to one order less than the highest-order derivative appearing in the integral formulation of the element equations must be continuous. Given the discretized representation of equation is given by:

$$\phi^{(e)}(x,y,z) = \sum_{i=1}^{M} N_i(x,y,z)\phi_i$$

It follows that the interpolation functions must meet this condition, since these functions determine the spatial variation of the field variable.

Recalling the application of Galerkin's method to the formulation of the truss element equations, the first derivative of the displacement appears in the equation (2). Therefore, the displacement must be continuous across element boundaries, but none of the displacement derivatives is required to be continuous across such boundaries. Indeed, as observed previously, the truss element is a constant strain element, so the first derivative is, in general, discontinuous at the boundaries.

Similarly, the beam element formulation, in the equation (3), includes the second derivative of displacement and compatibility requires continuity of both the displacement and the slope at the element boundaries. In addition to satisfying the criteria for convergence, the compatibility condition can be given a physical meaning as well.

In structural problems, the requirement of displacement continuity along element boundaries ensures that no gaps or voids develop in the structure as a result of modeling procedure. Similarly the requirement of slope continuity for the beam element ensures that no "kinks" are developed in the deformed structure. In heat transfer problems, the compatibility requirement prevents the physically unacceptable possibility of jump discontinuities in temperature distribution.

Compatibility Equation

$$\delta = \frac{PL}{EA}$$

Where,

P-Load

L-Length of original member

E-Modules of elasticity

A-Cross section area

$$\delta_T = \alpha_T \Delta T L$$

Where,

α_T-Linear coefficient of expression

ΔT-Change in temperature

L -Length of original member

δ_T -Change in length of number

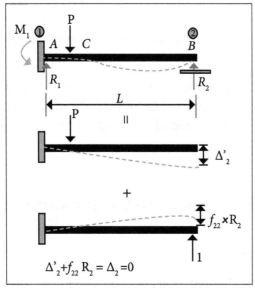

(a) Compatibility of displacement.

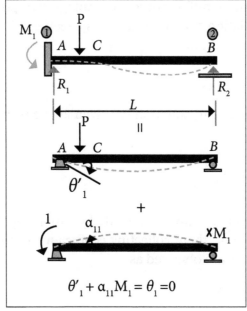

(b) Compatibility of slope.

Completeness

In the limit as element size shrinks to zero in mesh refinement, the field variable and its partial derivatives up to and including, the highest-order derivative appearing in the integral formulation must be capable of assuming constant values. Again, because of the Discretization, the completeness requirement is directly applicable to the interpolation functions.

The completeness requirement ensures that a displacement field within a structural element can take on a constant value, representing rigid body motion, for example. Similarly, constant slope of a beam element represents rigid body rotation, while a state of constant temperature in a thermal element corresponds to no heat flux through the element.

In addition to the rigid body motion consideration, the completeness requirement also ensures the possibility of constant values of (at least) first derivatives. This feature assures that a finite element is capable of constant strain, constant heat flow or constant fluid velocity, for examples.

3.2 Polynomial Forms: One Dimensional Element

One Dimensional Element

The formulation of finite element characteristics requires differentiation and integration of the interpolation functions in various forms. Owing to the simplicity with which polynomial functions can be differentiated and integrated, polynomials are the most commonly used interpolation functions. The displacement field is expressed via the first-degree polynomial equation:

$$u(x) = a_0 + a_1 x \qquad \qquad ...(1)$$

In terms of nodal displacement, equation (2) given above is determined to be equivalent to:

$$u(x) = \left(1 - \frac{x}{L}\right) u_1 + \frac{x}{L} u_2 \qquad \qquad ...(2)$$

The coefficients a0 and a1 are obtained by applying the nodal conditions $u(x=0) = u_1$ and $u(x=L) = u_2$. then; collecting coefficients of the nodal displacements, the interpolation functions are obtained as:

$$N_1 = 1 - \frac{x}{L} \qquad N_2 = \frac{x}{L} \qquad \qquad ...(3)$$

The equation (3) shows that, if $u_1 = u_2$. The element displacement field corresponds to rigid body motion and no straining of the element occurs.

The first derivative of equation (3) with respect to x yields a constant value. Hence, the truss element satisfies the completeness requirement, since both displacement and strain can take on constant values regardless of element size. Also note that the truss element satisfies the compatibility requirement automatically, since only displacement is involved and displacement compatibility is enforced at the nodal connections via the system assembly procedure.

In light of the completeness requirement, we can now see that choice of the linear polynomial representation of the displacement field, equation (1), was not arbitrary. Inclusion of the constant term a0 ensures the possibility of rigid body motion, while the first-order term provides for a constant first derivative.

Further, only two terms can be included in the representation, as only two boundary conditions have to be satisfied, corresponding to the two element degrees of freedom. Conversely, if the linear term were to be replaced by a quadratic term a_2x^2, for example.

The coefficients could still be obtained to mathematically satisfy the nodal displacement conditions, but constant first derivative could not be obtained under any circumstances.

Determination of the interpolation functions for the truss element. Nevertheless, the procedure is typical of that used to determine the interpolation functions for any element in which polynomials are utilized. We revisit the development of the beam element interpolation functions with specific reference to the compatibility and completeness requirements.

By including the slopes at element nodes as nodal variables in addition to nodal displacements, the compatibility condition is satisfied via the system assembly procedure. The beam element then has 4°of freedom and the displacement field is represented as the cubic polynomial:

$$v(x) = a_0 + a_1x + a_2x^2 + a_3x^3 \qquad \qquad ...(4)$$

Which is ultimate by the expressed in terms of interpolation function and nodal variable as:

$$v(x) = N_1v_1 + N_2\theta_1 + N_3v_2 + N_4\theta_2 = \begin{bmatrix} N_1 & N_2 & N_3 & N_4 \end{bmatrix} \begin{Bmatrix} v_1 \\ \theta_1 \\ v_2 \\ \theta_2 \end{Bmatrix} \qquad ...(5)$$

Rewriting equation (4) as the matrix product:

$$v(x) = \begin{bmatrix} 1 & x & x^2 & x^3 \end{bmatrix} \begin{Bmatrix} a_0 \\ a_1 \\ a_2 \\ a_3 \end{Bmatrix} \qquad \qquad ...(6)$$

The nodal conditions:

$$v(x = 0) = v_1$$

$$\left.\frac{dv}{dx}\right|_{x=0} = \theta_1$$

$$v(x = L) = v_2$$

$$\left.\frac{dv}{dx}\right|_{x=L} = \theta_2 \qquad \qquad ...(7)$$

Are applied to obtain:

$$v_1 = \begin{bmatrix} 1 & 0 & 0 & 0 \end{bmatrix} \begin{Bmatrix} a_0 \\ a_1 \\ a_2 \\ a_3 \end{Bmatrix} \qquad \qquad ...(8)$$

$$\theta_1 = \begin{bmatrix} 0 & 1 & 0 & 0 \end{bmatrix} \begin{Bmatrix} a_0 \\ a_1 \\ a_2 \\ a_3 \end{Bmatrix} \qquad \qquad ...(9)$$

$$v_2 = \begin{bmatrix} 1 & L & L^2 & L^3 \end{bmatrix} \begin{Bmatrix} a_0 \\ a_1 \\ a_2 \\ a_3 \end{Bmatrix} \qquad \qquad ...(10)$$

$$\theta_2 = \begin{bmatrix} 0 & 1 & 2L & 3L^2 \end{bmatrix} \begin{Bmatrix} a_0 \\ a_1 \\ a_2 \\ a_3 \end{Bmatrix} \qquad \qquad ...(11)$$

The last four equations are combined into the equivalent matrix form:

$$\begin{Bmatrix} v_1 \\ \theta_1 \\ v_2 \\ \theta_2 \end{Bmatrix} = \begin{bmatrix} 1 & 0 & 0 & 0 \\ 0 & 1 & 0 & 0 \\ 1 & L & L^2 & L^3 \\ 0 & 1 & 2L & 3L^2 \end{bmatrix} \begin{Bmatrix} a_0 \\ a_1 \\ a_2 \\ a_3 \end{Bmatrix} \qquad \qquad ...(12)$$

The system represented by equation (12) can be solved for the polynomial coefficients by inverting the coefficient matrix to obtain:

$$
\begin{Bmatrix} a_0 \\ a_1 \\ a_2 \\ a_3 \end{Bmatrix} =
\begin{bmatrix}
1 & 0 & 0 & 0 \\
0 & 1 & 0 & 0 \\
-\dfrac{3}{L^2} & -\dfrac{2}{L} & \dfrac{3}{L^2} & -\dfrac{1}{L} \\
\dfrac{2}{L^3} & \dfrac{1}{L^2} & -\dfrac{2}{L^3} & \dfrac{1}{L^2}
\end{bmatrix}
\begin{Bmatrix} v_1 \\ \theta_1 \\ v_2 \\ \theta_2 \end{Bmatrix}
\qquad \ldots(13)
$$

The interpolation functions can now be obtained by substituting the coefficients given by equation (13) into equation (4) and collecting coefficients of the nodal variables. However, the following approach is more direct and algebraically simpler. Substitute equation (13) into equation (6) and (5) to obtain:

$$
v(x) = \begin{bmatrix} 1 & x & x^2 & x^3 \end{bmatrix}
\begin{bmatrix}
1 & 0 & 0 & 0 \\
0 & 1 & 0 & 0 \\
-\dfrac{3}{L^2} & -\dfrac{2}{L} & \dfrac{3}{L^2} & -\dfrac{1}{L} \\
\dfrac{2}{L^3} & \dfrac{1}{L^2} & -\dfrac{2}{L^3} & \dfrac{1}{L^2}
\end{bmatrix}
\begin{Bmatrix} v_1 \\ \theta_1 \\ v_2 \\ \theta_2 \end{Bmatrix}
$$

$$
= \begin{bmatrix} N_1 & N_2 & N_3 & N_4 \end{bmatrix}
\begin{Bmatrix} v_1 \\ \theta_1 \\ v_2 \\ \theta_2 \end{Bmatrix}
\qquad \ldots(14)
$$

The interpolation functions are:

$$
\begin{bmatrix} N_1 & N_2 & N_3 & N_4 \end{bmatrix} = \begin{bmatrix} 1 & x & x^2 & x^3 \end{bmatrix}
$$

$$
\begin{bmatrix}
1 & 0 & 0 & 0 \\
0 & 1 & 0 & 0 \\
-\dfrac{3}{L^2} & -\dfrac{2}{L} & \dfrac{3}{L^2} & -\dfrac{1}{L} \\
\dfrac{2}{L^3} & \dfrac{1}{L^2} & -\dfrac{2}{L^3} & \dfrac{1}{L^2}
\end{bmatrix}
\qquad \ldots(15)
$$

$$
P_M(x,y) = \sum_{k=0}^{N(2)} a_k x^i y^j \qquad i+j \le M \qquad \ldots(16)
$$

And note that the results of equation (15) are identical to those shown in equation (16).

The Development of the Beam Element Interpolation Function

- To establish a general procedure for use with polynomial representations of the field variable.

- To revisit the beam clement formulation in terms of compatibility and completeness requirements.

- The general procedure begins with expressing the field variable as a polynomial of order one fewer than the number of degrees of freedom exhibited by the element. Using the examples of the truss and beam elements, it has been shown that a two-node element may have 2 degrees of freedom, as in the truss element where only displacement continuity is required or 4 degrees of freedom, as in the beam element where slope continuity is required.

- Next the nodal (boundary) conditions are applied and the coefficients of the polynomial are computed accordingly. Finally, the polynomial coefficients are substituted into the field variable representation in terms of nodal variables to obtain the explicit form of the interpolation functions.

3.2.1 Geometric Isotropy

The polynomial representation of the field variable must contain the same number of terms as the number of nodal degrees of freedom. In addition, to satisfy the completeness requirement, the polynomial representation for an M-degree of freedom element should contain all powers of the independent variable up to and including M-1. Another way of stating the latter requirement is that the polynomial is complete.

In two and three dimensions, polynomial representations of the field variable, in general, satisfy the compatibility and completeness requirements if the polynomial exhibits the property known as geometric isotropy. A mathematical function satisfies geometric isotropy if the functional form does not change under a translation or rotation of coordinates. In two dimensions, a complete polynomial of order M can be expressed as:

$$P_M(x,y)=\sum_{k=0}^{N(2)}a_k x^i y^j \qquad i+j\leq M \qquad ...(1)$$

Where, $N_t^{(2)} = \lceil(M+1)(M+2)\rceil/2$ is the total number of terms. A complete polynomial as expressed by equation (1) satisfies the condition of geometric isotropy, since the two variables, x and y, are included in each term in similar powers. Therefore, a translation or rotation of coordinates is not prejudicial to either independent variable.

A graphical method of depicting complete two-dimensional polynomials is called Pascal triangle, shown in the figure (a). Each horizontal line represents a polynomial of order M. A complete polynomial of order M must contain all terms shown above the horizontal line. For example, a complete quadratic polynomial in two dimensions must contain six terms. Hence, for use in a finite element representation of a field variable, a complete quadratic expression requires six nodal degrees of freedom in the element.

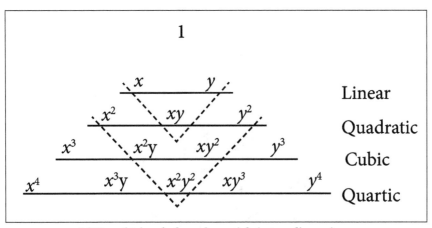

(a) Pascal triangle for polynomials in two dimensions.

In addition to the complete polynomials, incomplete polynomials also exhibit geometric isotropy if the incomplete polynomial is symmetric. In this context, symmetry implies that the independent variables appear as "equal and opposite pairs," ensuring that each independent variable plays an equal role in the polynomial.

$$P(x,y) = a_0 + a_1x + a_2y + a_3x^2 \qquad\qquad ...(2)$$

For example, the four-term incomplete quadratic polynomial is not symmetric, as there is a quadratic term in x but the corresponding quadratic term in y does not appear. On the other hand, the incomplete quadratic polynomial is symmetric, as the quadratic term gives equal "weight" to both variables.

$$P(x,y) = a_0 + a_1x + a_2y + a_3xy \qquad\qquad ...(3)$$

A very convenient way of visualizing some of the commonly used incomplete but symmetric polynomials of a given order is also afforded by the Pascal triangle. Again referring to figure (a) the dashed lines show the terms that must be included in an incomplete yet symmetric polynomial of a given order. All terms above the dashed lines must be included in a polynomial representation if the function is to exhibit geometric isotropy.

This feature of polynomials is utilized to a significant extent in following the development of various element interpolation functions.

As in the two-dimensional case, to satisfy the geometric requirements, the polynomial expression of the field variable in three dimensions must be complete or incomplete but symmetric. Completeness and symmetry can also be depicted graphically by the "Pascal pyramid" shown in the figure (b). While the three-dimensional case is a bit more difficult to visualize, the basic premise remains that each independent variable must be of equal "strength" in the polynomial.

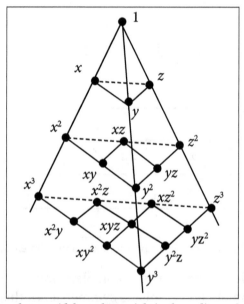

Pascal pyramid for polynomials in three dimension.

For example, the 3-D quadratic polynomial is complete and could be applied to an element having 10 nodes.

$$P(x,y,z)=a_0+a_1x+a_2y+a_3z+a_4x^2+a_5y^2+a_6z^2$$

$$+a_7xy+a_8xz+a_9yz \qquad ...(4)$$

Similarly, an incomplete, symmetric form such as:

$$P(x,y,z)=a_0+a_1x+a_2y+a_3z+a_4x^2+a_5y^2+a_6z^2 \qquad ...(5)$$

Or

$$P(x,y,z)=a_0+a_1x+a_2y+a_3z+a_4xy+a_5xz+a_6yz \qquad ...(6)$$

Could be used for elements having seven nodal degrees of freedom (an unlikely case, however).

Geometric isotropy is not an absolute requirement for field variable representation, hence, interpolation functions. As demonstrated by many researchers, incomplete representations are quite often used and solution convergence attained. However, in terms of h-refinement, use of geometrically isotropic representations guarantees satisfaction of the compatibility and completeness requirements.

For the p-refinement method is interpolation functions in any finite element analysis solution are approximations to the power series expansion of the problem solution. As we increase the number of element nodes, the order of the interpolation functions increases and, in the limit, as the number of nodes approaches infinity, the polynomial expression of the field variable approaches the power series expansion of the solution.

3.2.2 Triangular Elements

Constant Strain Triangular (CST) Element

A three noded triangular element is known as Constant Strain Triangular element (CST) which is shown in figure below. It has six unknown displacement degrees of freedom (u_1v_1, u_2v_2, u_3v_3). The element is called CST because it has a constant strain throughout it.

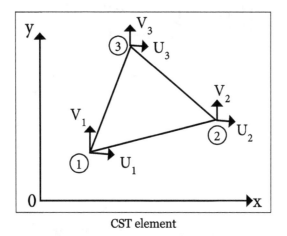

CST element

When we develop an element, we start by stipulating a displacement model in the form of a polynomial or trigonometric function.

Functions are:

$$u(x, y) = a_1 + a_2x + a_3y$$

$$v(x, y) = a_4 + a_5x + a_6y$$

$$\varepsilon_x = \frac{\partial v}{\partial y} = a_2, \text{constant.}$$

$$\varepsilon_y = \frac{\partial v}{\partial y} = a_6, \text{constant.}$$

$$\gamma_{xy} = \frac{\partial v}{\partial y} + \frac{\partial v}{\partial y} = a_2 + a_6, \text{constant.}$$

This will lead to a situation where, σ_x, σ_y and τ_{xy}

Will take some constant values inside the element.

CST element is the simplest form of 2D elements. We get the stiffness matrix uniquely in terms of the geometry of the element and the elasticity matrix.

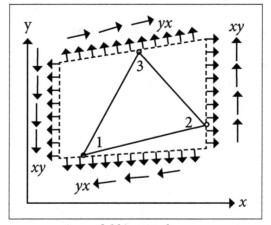

Stress field in CST element.

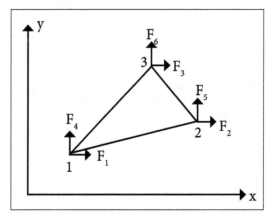

Equivalent Nodal Forces.

$$F_1 = \frac{1}{2} \left[\sigma_x (y_2 - y_3) + \tau_{xy} (x_3 - x_2) \right]$$

$$F_2 = \frac{1}{2} \left[\sigma_x (y_3 - y_1) + \tau_{xy} (x_1 - x_3) \right]$$

$$F_3 = \frac{4}{2}\left[\sigma_x\left(y_1 - y_2\right) + \tau_{xy}\left(x_2 - x_1\right)\right]$$

$$F_4 = \frac{t}{2}\left[\sigma_y\left(x_3 - x_2\right) + \tau_{xy}\left(y_2 - y_3\right)\right]$$

$$F_5 = \frac{t}{2}\left[\sigma_y\left(x_1 - x_3\right) + \tau_{xy}\left(y_3 - y_1\right)\right]$$

$$F_6 = \frac{t}{2}\left[\sigma_y\left(x_2 - x_1\right) + \tau_{xy}\left(y_1 - y_2\right)\right]$$

Linear Strain Triangle (LST) Element

A six noded triangular element is known as Linear Strain triangular (LST) element which is shown in figure below. It has twelve unknown displacement degrees of freedom. The displacement functions of the element are quadratic instead of linear as in the CST.

The procedures for development of the stiffness matrix equations for the LST element follow the same steps as those used for the CST element. But the number of equations used for developing no shift matrix equation is 12 instead of 6. It is a tedious process to solve those equations. Hence, we will use a computer to solve many of the mathematical equations.

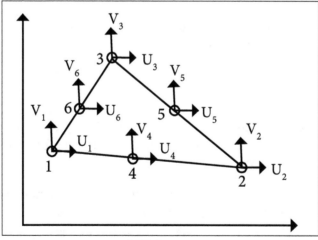

LST element

LST element is preferred than the CST element for plane stress applications when relatively small numbers of nodes are used. LST element is not preferred when large numbers of nodes are used since, the cost of formation of the element stiffness equation band width are high compared to CST element. Computer modeling for large number of nodes are also difficult to LST element.

Shape Functions for LST:

$$N_1 = L_1(2L_1 - 1)$$

$$N_2 = L_2(2L_2 - 1)$$

$$N_3 = L_3(2L_3 - 1)$$

$$N_4 = 4\,L_1 L_2$$

$$N_5 = 4\,L_2 L_3$$

$$N_6 = 4\,L_3 L_1$$

$$u_1 = N_1 u_1 + N_2 u_2 + N_3 u_3 + N_3 u_4 + N_5 u_5 + N_6 u_6$$

$$\varepsilon_x = \frac{\partial u}{\partial x}\left[\frac{dN_1}{dx}\;\frac{dN_2}{dx}\;\frac{dN_3}{dx}\cdot\;\frac{dN_4}{dx}\;\frac{dN_5}{dx}\;\frac{dN_6}{dx}\right]\begin{Bmatrix} u_2 \\ u_2 \\ \cdot \\ \cdot \\ \cdot \\ \cdot \\ u_6 \end{Bmatrix}$$

$$\frac{dN_1}{dx} = \frac{dN_1}{dL_1}\frac{dN_1}{dx} + \frac{dN_1}{dL_2}\frac{dN_2}{dx} + \frac{dN_1}{dL_3}\frac{dN_3}{dx}$$

3.2.3 Rectangular Elements

Rectangular elements are convenient for use in modeling regular geometries, can be used in conjunction with triangular elements and form the basis for development of general quadrilateral elements.

The simplest of the rectangular family of elements is the four-node rectangle shown in the figure (a), where it is assumed that the sides of the rectangular are parallel to the global Cartesian axes. By convention, we number the nodes sequentially in a counter-clockwise direction, as shown.

As there are four nodes and 4 degrees of freedom, a four-term polynomial expression for the field variable is appropriate. Since there is no complete four-term polynomial in two dimensions, the incomplete, symmetric expression is used to ensure geometric isotropy.

$$\phi(x,y) = a_0 + a_1 x + a_2 y + a_3 xy \qquad\qquad ...(1)$$

Applying the four nodal conditions and writing in matrix form gives:

$$\begin{Bmatrix} \phi_1 \\ \phi_2 \\ \phi_3 \\ \phi_4 \end{Bmatrix} = \begin{bmatrix} 1 & x_1 & y_1 & x_1y_1 \\ 1 & x_3 & y_3 & x_2y_2 \\ 1 & x_3 & y_3 & x_3y_3 \\ 1 & x_4 & y_4 & x_4y_4 \end{bmatrix} \begin{Bmatrix} a_0 \\ a_1 \\ a_2 \\ a_3 \end{Bmatrix} \qquad \text{...}(2)$$

which formally gives the polynomial coefficients as:

$$\begin{Bmatrix} a_0 \\ a_1 \\ a_2 \\ a_3 \end{Bmatrix} = \begin{bmatrix} 1 & x_1 & y_1 & x_1y_1 \\ 1 & x_3 & y_3 & x_2y_2 \\ 1 & x_3 & y_3 & x_3y_3 \\ 1 & x_4 & y_4 & x_4y_4 \end{bmatrix}^{-1} \begin{Bmatrix} \phi_1 \\ \phi_2 \\ \phi_3 \\ \phi_4 \end{Bmatrix} \qquad \text{...}(3)$$

In terms of the nodal values, the field variable is then described by:

$$\phi(x,y) = \begin{bmatrix} 1 & x & y & xy \end{bmatrix} \begin{bmatrix} 1 & x_1 & y_1 & x_1y_1 \\ 1 & x_3 & y_3 & x_2y_2 \\ 1 & x_3 & y_3 & x_3y_3 \\ 1 & x_4 & y_4 & x_4y_4 \end{bmatrix}^{-1} \begin{Bmatrix} \phi_1 \\ \phi_2 \\ \phi_3 \\ \phi_4 \end{Bmatrix} \qquad \text{...}(4)$$

From which the interpolation functions can be deduced.

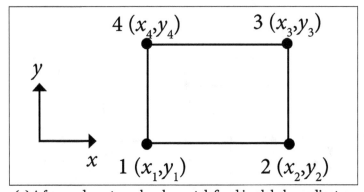

(a) A four node rectangular element defined in global coordinates.

The form of equation (4) suggests that expression of the interpolation functions in terms of the nodal coordinates is algebraically complex. Fortunately, the complexity can be reduced by a more judicious choice of coordinates. For the rectangular element, we introduce the normalized coordinates (also known as natural coordinates or serendipity coordinates) r and s as:

$$r = \frac{x - \bar{x}}{a} \qquad\qquad s = \frac{y - \bar{y}}{b} \qquad \text{...}(5)$$

Where, 2a and 2b are the width and height of the rectangle, respectively and the coordinates of the centroid are:

$$\bar{x}=\frac{x_1+x_2}{2} \qquad\qquad \bar{y}=\frac{y_1+y_4}{2} \qquad\qquad\qquad ...(6)$$

As shown in the figure (b). Therefore, r and s are such that the values range from -1 to +1 and the nodal coordinates are as in the figure (c). Applying the conditions that must be satisfied by each interpolation function at each node, we obtain (essentially by inspection).

$$N_1(r,s)=\frac{1}{4}(1-r)(1-s)$$

$$N_2(r,s)=\frac{1}{4}(1+r)(1-s)$$

$$N_3(r,s)=\frac{1}{4}(1+r)(1+s)$$

$$N_4(r,s)=\frac{1}{4}(1-r)(1+s) \qquad\qquad\qquad ...(7)$$

Hence,

$$\phi(x,y)=\phi(r,s)=N_1(r,s)\phi_1+N_2(r,s)\phi_2$$

$$+N_3(r,s)\phi_3+N_4(r,s)\phi_4 \qquad\qquad\qquad ....(8)$$

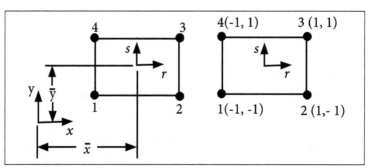

(b) Natural coordinates (c) Natural coordinates of each node.

As in the case of triangular elements using area coordinates, the interpolation functions are much simpler algebraically when expressed in terms of the natural coordinates. Nevertheless, all required conditions are satisfied and the functional form is identical to that used to express the field variable in the equation (1).

Also as with area coordinates, integrations involving the interpolation functions expressed in the natural coordinates are simplified, since the integrands are relatively simple polynomials (for rectangular elements) and the integration limits are -1 and + 1. Further discussion of such integration requirements, particularly numerical integration techniques, is postponed until later in this chapter.

To develop a higher-order rectangular element, the logical progression is to place an additional node at the midpoint of each side of the element, as in the figure (d).

This proses an immediate problem, however. Inspection of the Pascal triangle shows that we cannot construct a complete polynomial having eight terms, but we have a choice of two incomplete, symmetric cubic polynomials.

$$\phi(x,y)=a_0+a_1x+a_2y+a_3x^2+a_4xy+a_5y^2+a_6x^3+a_7y^3$$

$$\phi(x,y)=a_0+a_1x+a_2y+a_3x^2+a_4xy+a_5y^2+a_6x^2y+a_7y^2$$

Rather than grapple with choosing one or the other, we use the natural coordinates and the nodal conditions that must be satisfied by each interpolation function to obtain the functions serendipitously. For example, interpolation function N1 must evaluate to zero at all nodes except node 1, where, its value must be unity. At nodes 2, 3 and 6, r = 1, so including the term r - 1 satisfies the zero condition at those nodes.

Similarly, at nodes 4 and 7, s = 1 so the term s -1 ensures the zero condition at those two nodes. Finally, at node 5, (r, s) = (0, -1) and at node 8, (r, s) = (-1, 0). Hence, at nodes 5 and 8, the term r + s + 1 is identically zero. Using this reasoning, the interpolation function associated with node 1 is to be of the form:

$$N_1(r,s)=(1-r)(1-s)(r+s+1) \qquad\qquad ...(9)$$

Evaluating at node 1 where, (r, s) = (-1.-1). We obtain N1 = -4, so a correction is required to obtain the unity value. The final form is then:

$$N_1(r,s)=\frac{1}{4}(r-1)(1-s)(r+s+1) \qquad\qquad ...(10)$$

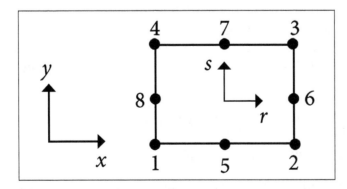

Eight-node rectangular element showing both global and natural coordinate axes.

A parallel procedure for the interpolation functions associated with the other three corner nodes leads to :

$$N_2(r,s)=\frac{1}{4}(r+1)(1-s)(s-r+1)$$

$$N_3(r,s)=\frac{1}{4}(1+r)(1+s)(r+s-1)$$

$$N_4(r,s)=\frac{1}{4}(r-1)(1+s)(r-s+1)$$

The form of the interpolation functions associated with the mid side nodes are simpler to obtain than those for the corner nodes. For example, N5 has a value of zero at nodes 2, 3 and 6 if it contains the term r -1 and is also zero at nodes 1, 4 and 8 if the term 1 + r is included. Finally, if a zero value is at node 7, (r, s) = (0, 1) is obtained by inclusion of s -1. The form for N5 is given by:

$$N_5=\frac{1}{2}(1-r)(1+r)(1-s)=\frac{1}{2}(1-r^2)(1-s)$$

Where, the leading coefficient ensures a unity value at node 5. For the other mid side nodes, are determined in the same manner.

$$N_6=\frac{1}{2}(1+r)(1-s^2)$$

$$N_7=\frac{1}{2}(1-r^2)(1+s)$$

$$N_8=\frac{1}{2}(1-r)(1-s^2)$$

Many other, successively higher-order, rectangular elements have been developed. In general, these higher-order elements include internal nodes that, in modeling, are troublesome, as they cannot be connected to nodes of other elements. The internal nodes are eliminated mathematically. The elimination process is such that the mechanical effects of the internal nodes are assigned appropriately to the external nodes.

3.2.4 Lagrange and Serendipity Elements

Lagrange Interpolation Function

An alternate and simpler way to derive shape functions is to use Lagrange interpolation polynomials. This method is suitable to derive shape function for elements having higher order of nodes. The Lagrange interpolation function at node 1 is defined by:

$$f_i(\xi)=\prod_{\substack{j=1\\j\neq i}}^{n}\frac{(\xi-\xi_j)}{(\xi_i-\xi_j)}=\frac{(\xi-\xi_1)(\xi-\xi_2)....(\xi-\xi_{i-1})(\xi-\xi_{i+1})....(\xi-\xi_n)}{(\xi_i-\xi_1)(\xi_i-\xi_2)....(\xi_i-\xi_{i-1})(\xi_i-\xi_{i+1})....(\xi_i-\xi_n)} \qquad ...(1)$$

The function $f_i(\xi)$ produces the Lagrange interpolation function for i_{th} node and ξ_j denotes ξ coordinate of j_{th} node in the element. In the above equation if we put $\xi = \xi_j$ and $j \neq i$, the value of the function $f_i(\xi)$ will be equal to zero.

Similarly, putting $\xi = \xi_i$, the numerator will be equal to denominator and hence $f_i(\xi)$ will have a value of unity. Since, Lagrange interpolation function for i_{th} node includes product of all terms except j_{th} term, for an element with nth nodes, $f_i(\xi)$ will have n-1 degrees of freedom. Thus, for one-dimensional elements with n-nodes we can define shape function as $N(\xi)\, f_i(\xi)$.

Shape Function for Two Node Bar Element

Let us consider the natural coordinate of the center of the element as 0 and the natural coordinate of the nodes 1 and 2 are -1 and +1 respectively. Therefore, the natural coordinate ξ at any point x can be represented by:

$$\xi = \frac{2(x-x_1)}{1}-1 \qquad ...(2)$$

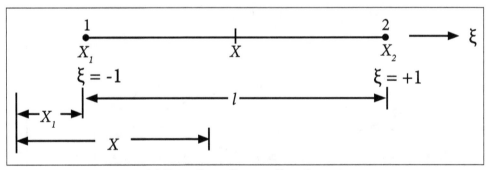

(a) Natural coordinates of bar element.

The shape function for two node bar element as shown in the figure (a), can be derived from equation (2) as follows:

$$N_1=f_1(\xi)=\frac{(\xi-\xi_2)}{(\xi_1-\xi_2)}=\frac{(\xi-1)}{-1-(1)}=\frac{1}{2}(1-\xi)$$

$$N_2=f_1(\xi)=\frac{(\xi-\xi_1)}{(\xi_2-\xi_1)}=\frac{(\xi+1)}{1-(1)}=\frac{1}{2}(1+\xi) \qquad ...(3)$$

Graphically, these shape functions are represented in the figure (b).

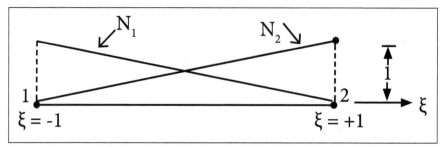

(b) Shape functions for two node bar element.

Shape Function for Three Node Bar Element

For a three node bar element as shown in the figure (c), the shape function will be quadratic in nature. These can be derived in the similar fashion using equation (1) which will be as follows:

$$N_1(\xi)=f_1(\xi)=\frac{(\xi-\xi_2)(\xi-\xi_3)}{(\xi_1-\xi_2)(\xi_1-\xi_3)}=\frac{(\xi)(\xi-1)}{(-1)(-2)}=\frac{1}{2}\xi(\xi-1)$$

$$N_2(\xi)=f_2(\xi)=\frac{(\xi-\xi_1)(\xi-\xi_3)}{(\xi_2-\xi_1)(\xi_2-\xi_3)}=\frac{(\xi+1)(\xi-1)}{(1)(-1)}=(1-\xi^2) \qquad ...(4)$$

$$N_3(\xi)=f_3(\xi)=\frac{(\xi-\xi_1)(\xi-\xi_2)}{(\xi_3-\xi_1)(\xi_3-\xi_2)}=\frac{(\xi+1)(\xi)}{(2)(1)}=\frac{1}{2}\xi(\xi+1)$$

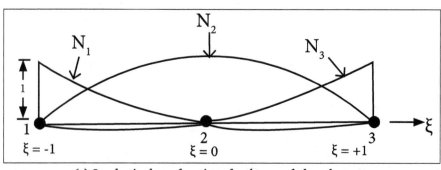

(c) Quadratic shape functions for three node bar element.

Shape Function for Two Dimensional Elements

We can derive the Lagrange interpolation function for two or three dimensional elements from one dimensional element as discussed above. Those elements whose shape functions are derived from the products of one dimensional Lagrange interpolation functions are called Lagrange elements.

The Lagrange interpolation function for a rectangular element can be obtained from the product of appropriate interpolation functions in the ξ direction $f_i(\xi)$ and η direction $[f_i(\eta)]$. Thus,

$$N_i(\xi,\eta) = f_i(\xi) = f_i(\eta) \qquad \qquad ...(5)$$

Where, i = 1, 2, 3, ..., n-node.

The procedure is described in details in following examples.

Four Node Rectangular Element

The shape functions for the four node rectangular element as shown in the figure (d) can be derived by applying equation (3) and (5) which will be as follows:

$$N_i(\xi,\eta) = f_i(\xi)f_i(\eta) = \frac{(\xi-\xi_2)}{(\xi_1-\xi_2)}\frac{(\eta-\eta_2)}{(\eta_1-\eta_2)} \qquad \qquad ...(6)$$

$$= \frac{(\xi-1)}{-1-(1)} \times \frac{(\eta-1)}{-1-(1)} = \frac{1}{4}(1-\xi)(1-\eta)$$

Similarly, other interpolation functions can be derived which are given below:

$$N_2(\xi,\eta) = f_2(\xi)f_2(\eta) = \frac{1}{4}(1+\xi)(1-\eta)$$

$$N_3(\xi,\eta) = f_2(\xi)f_2(\eta) = \frac{1}{4}(1+\xi)(1-\eta)$$

$$N_4(\xi,\eta) = f_1(\xi)f_2(\eta) = \frac{1}{4}(1+\xi)(1-\eta) \qquad \qquad ...(7)$$

which was derived earlier by choosing polynomials.

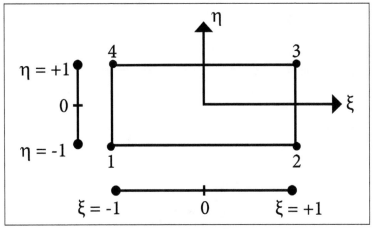

(d) Four node rectangular element.

Nine Node Rectangular Element

In a similar way, to the derivation of four node rectangular element, we can derive the shape functions for a nine node rectangular element. In this case, the shape functions can be derived using equation (4) and (5).

$$N_1(\xi,\eta) = f_1(\xi)f_1(\eta) = \frac{1}{2}\xi(\xi-1) \times \frac{1}{2}\eta(\eta-1) = \frac{1}{4}\xi\eta(\xi-1)(\eta-1) \quad \ldots(8)$$

In a similar way, all the other shape functions of the element can be derived. The shape functions of nine node rectangular element will be:

$$N_1 = \frac{1}{4}\xi\eta(\xi-1)(\eta-1), \qquad N_2 = \frac{1}{4}\xi\eta(\xi+1)(\eta-1)$$

$$N_3 = \frac{1}{4}\xi\eta(\xi+1)(\eta+1) \qquad N_4 = \frac{1}{4}\xi\eta(\xi-1)(\eta+1)$$

$$N_5 = \frac{1}{2}\eta(1-\xi^2)(\eta-1) \qquad N_6 = \frac{1}{2}\xi(\xi+1)(1-\eta^2) \qquad \ldots(9)$$

$$N_7 = \frac{1}{2}\eta(1-\xi^2)(\eta+1) \qquad N_8 = \frac{1}{2}\xi(\xi-1)(1-\eta^2)$$

$$N_9 = (1-\xi^2)(1-\eta^2)$$

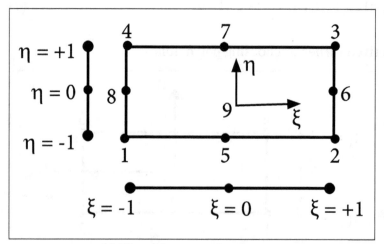

(e) Nine node rectangular element.

Thus, it is observed that the two dimensional Lagrange element contains internal nodes (figure (f)) which are not connected to other nodes.

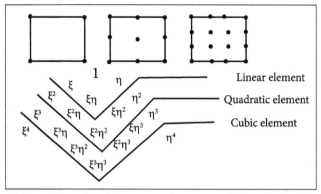

(f) Two dimensional Lagrange elements and Pascal triangle.

Serendipity Elements

The serendipity family elements in figure below, these elements may be called as boundary node family elements also. In these elements nodes are only on the boundaries. Sienkiewicz called them as 'Serendip family' elements by referring to the famous princess of Serendip noted for chance discoveries. The terms linear, quadratic, cubic and quartic are used since the variation of shape functions about a boundary is of that order.

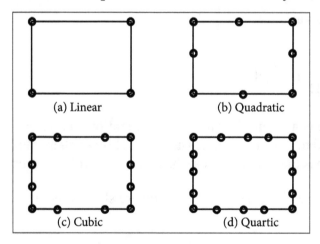

(a) Linear   (b) Quadratic

(c) Cubic   (d) Quartic

(A) Linear Element:

$$N_i(\xi,\eta)=\frac{1}{4}(1+\xi\xi_i)(1+\eta\eta_i)$$

(B) Quadratic Element:

(i) For nodes at $\xi=\pm1, \eta=\pm1$

$$N_i(\xi,\eta)=\frac{1}{4}(1+\xi\xi_i)(1+\eta\eta_i)(\xi\xi_i+\eta\eta_i-1)$$

(ii) For nodes at $\xi=\pm1, \eta=0$

$$N_i(\xi,\eta)=\frac{1}{2}(1+\xi\xi_i)(1+\eta^2)$$

(iii) For nodes at $\xi=0, \eta=\pm1$

$$N_i(\xi,\eta)=\frac{1}{2}(1+\xi^2)(1+\eta\eta_i)$$

(C) Cubic Element:

(i) For nodes at $\xi=\pm1, \eta=\pm1$

$$N_i(\xi,\eta)=\frac{1}{32}(1+\xi\xi_i)(1+\eta\eta_i)\left[9(\xi^2+\eta^2)-10\right]$$

(ii) For nodes at $\xi=\pm1, \eta=\pm\frac{1}{3}$

$$N_i(\xi,\eta)=\frac{9}{32}(1+\xi\xi_i)(1-\eta^2)(1+9\eta\eta_i)$$

The shape functions are found from the consideration that Ni = 1 for ith node and is zero when referred to any other node. Discovery of these elements clubbed with Isoperimetric concept has made major breakthrough in the finite element analysis.

Example: Using 'serendipity concept' let us derive shape functions for 4 noded rectangular elements.

Solution:

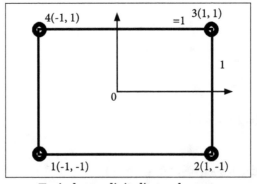

Typical serendipity linear element.

N_1 has to satisfy the conditions,

(a) along $\xi = 1, N_1 = 0$.

(b) along $\eta = 1, N_1 = 0$.

(c) at $\xi = -1, \eta = 1, N_1 = 0$.

Hence let,

$N_1 = C(1-\xi)(1-\eta)$, where C is arbitrary constant. Conditions (a) and (b) are satisfied. Condition (c) gives:

$$1 = C(1+1)(1+1) \qquad (\text{or}) \qquad C = \frac{1}{4}$$

$$\therefore N_1 = \frac{(1-\xi)(1-\eta)}{4}$$

On the same lines we can get:

$$N_2 = \frac{(1-\xi)(1-\eta)}{4}$$

$$N_3 = \frac{(1+\xi)(1+\eta)}{4}$$

$$\text{And, } N_4 = \frac{(1-\xi)(1-\eta)}{4}$$

3.2.5 Solid Elements

There are two basic families of three-dimensional elements similar to two-dimensional case.

Extension of triangular elements will produce tetrahedrons in three dimensions. Similarly, rectangular parallelepipeds are generated on the extension of rectangular elements. Shown in the figure below, a few commonly used solid elements for finite element analysis.

Derivation of shape functions for such three dimensional elements in Cartesian coordinates are algebraically quite cumbersome. This is observed while developing shape functions in two dimensions. Therefore, the shape functions for the two basic elements of the tetrahedral and parallelepipeds families will be derived using natural coordinates.

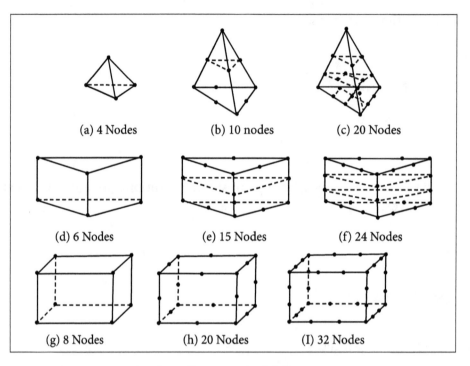

(a) Three-dimensional solid elements.

The polynomial expression of the field variable in three dimensions must be complete or incomplete but symmetric to satisfy the geometric isotropy requirements. Completeness and symmetry can be ensured using the Pascal pyramid which is shown in the figure (b). It is important to note that each independent variable must be of equal strength in the polynomial.

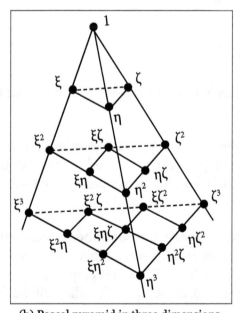

(b) Pascal pyramid in three dimensions.

The following 3-D quadratic polynomial with complete terms can be applied to an element having 10 nodes:

$$\phi(\xi,\eta,\zeta)=\alpha_0+\alpha_1\xi+\alpha_2\eta+\alpha_3\zeta+\alpha_4\xi^2+\alpha_5\eta^2+\alpha_6\zeta^2+\alpha_7\xi\eta+\alpha_8\eta\zeta+\alpha_9\zeta\xi$$

However, the geometric isotropy is not an absolute requirement for field variable representation to derive the shape functions.

Tetrahedral Elements

The simplest element of the tetrahedral family is a four node tetrahedron as shown in the figure (c). The node numbering has been followed in sequential manner, i.e., in this case anti-clockwise direction.

Similar to the area coordinates, the concept of volume co-ordinates has been introduced here. The coordinates of the nodes are defined both in Cartesian and volume coordinates. Point P(x, y and z) as shown in the figure (b), is an arbitrary point in the tetrahedron.

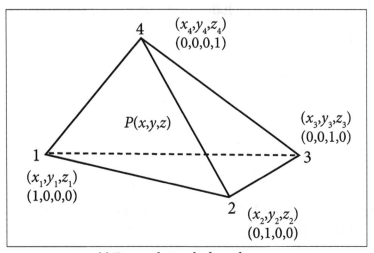

(c) Four node tetrahedron element.

The linear shape function for this element can be expressed as:

$$\{N\}=\begin{bmatrix} L_1 & L_2 & L_3 & L_4 \end{bmatrix}$$

Here, L_1 L_2 L_3 L_4 are the set of natural coordinates inside the tetrahedron and are defined as follows:

$$L_i=\frac{V_i}{V}$$

Where, V_i is the volume of the sub element which is bound by point P and face i and V is the total volume of the element. For example L_1 may be interpreted as the ratio of the volume of the sub element P234 to the total volume of the element 1234. The volume of the element V is given by the determinant of the nodal coordinates as follows:

$$V = \frac{1}{6}\begin{vmatrix} 1 & 1 & 1 & 1 \\ x_1 & x_2 & x_3 & x_4 \\ y_1 & y_2 & y_3 & y_4 \\ z_1 & z_2 & z_3 & z_4 \end{vmatrix}$$

The relationship between the Cartesian and natural coordinates of point P may be expressed as:

$$\begin{Bmatrix} 1 \\ x \\ y \\ z \end{Bmatrix} = \begin{bmatrix} 1 & 1 & 1 & 1 \\ x_1 & x_2 & x_3 & x_4 \\ y_1 & y_2 & y_3 & y_4 \\ z_1 & z_2 & z_3 & z_4 \end{bmatrix} \begin{Bmatrix} L_1 \\ L_2 \\ L_3 \\ L_4 \end{Bmatrix}$$

It may be noted that the identity included in the first row ensure the matrix invertible:

$$L_1 + L_2 + L_3 + L_4$$

The inverse relation is given by:

$$\begin{Bmatrix} L_1 \\ L_2 \\ L_3 \\ L_4 \end{Bmatrix} = \frac{1}{6V} \begin{bmatrix} V_1 & a_1 & b_1 & c_1 \\ V_2 & a_3 & b_3 & c_3 \\ V_3 & a_3 & b_3 & c_3 \\ V_4 & a_4 & b_4 & c_4 \end{bmatrix} \begin{Bmatrix} 1 \\ x \\ y \\ z \end{Bmatrix}$$

Here ,Vi is the volume subtended from face i and terms ai, bi and ci represent the projected area of face i on the x, y ,z coordinate planes respectively and are given as follows:

$$a_i = (z_j y_k - z_k y_j) + (z_k y_1 - z_1 y_k) + (z_1 y_j - z_j y_1)$$

$$b_i = (z_j x_k - z_k x_j) + (z_k x_1 - z_1 x_k) + (z_1 x_j - z_j x_1)$$

$$c_i = (y_j x_k - y_k x_j) + (y_k x_1 - y_1 x_k) + (y_1 x_j - y_j x_1)$$

i, j, k, l will be in cyclic order (i.e., 1→2→3→4→1). The volume coordinates full fill all nodal conditions for interpolation functions. Therefore, the field variable can be expressed in terms of nodal values as:

$$\phi(x,y,z) = L_1 \phi_1 + L_2 \phi_2 + L_3 \phi_3 + L_4 \phi_4$$

Though the shape functions (i.e., the volume coordinates) in terms of global coordinates is algebraically complex but they are straight forward. The partial derivatives of the natural coordinates with respect to the Cartesian coordinates are given by:

$$\frac{\partial L_i}{\partial x} = \frac{a_i}{6V}, \frac{\partial L_i}{\partial y} = \frac{b_i}{6V}, \frac{\partial L_i}{\partial z} = \frac{c_i}{6V}$$

Similar to area integral, the general integral taken over the volume of the element is given by:

$$\int_v L_1^p L_2^q L_3^r L_4^s \, dV = \frac{p!q!r!s!}{(p+q+r+s+3)!} \cdot 6V$$

The four node tetrahedral element is a linear function of the Cartesian coordinates. Hence, all the first partial derivatives of the field variable will be constant. The tetrahedral element is a constant strain element as the element exhibits constant gradients of the field variable in the coordinate directions.

Higher order elements of the tetrahedral family are shown in the figure (A). The shape functions for such higher order three dimensional elements can readily be derived in volume coordinates, as for higher-order two-dimensional triangular elements. The second element of this family has 10 nodes and a cubic form for the field variable and interpolation functions.

3.3 Isoparametric Formulation

The analysis of structural problems of complex shapes involving curved boundaries or surfaces, simple triangular or rectangular elements are no longer sufficient. These have led to the development of elements of more arbitrary shape and are called Isoparametric elements. These elements are widely used in two and three-dimensional stress analysis and, plates and shell problems.

The concept of Isoparametric element is based on the transformation of the parent element in local or natural coordinate system to an arbitrary shape in the Cartesian coordinate system as shown by examples in shown in the figure below. A convenient way of expressing the transformation is to make use of the shape functions of the rectilinear elements in their natural coordinate system and the nodal values of the co-ordinates. Thus the Cartesian coordinates of a point in an element may be expressed as:

$$x = N_1' \, x_1 + N_2' \, x_2 + \cdots + N_n' \, x_n$$

$$y = N_1' \, y_1 + N_2' \, y_2 + \cdots + N_n' \, y_n$$

$$z = N_1' \, z_1 + N_2' \, z_2 + \cdots + N_n' \, z_n$$

Or in matrix form:

$$\{x\} = [N']\{x_n\}$$

Where, $[N']$ are the shape functions of the parent rectilinear element and $\{x_n\}$ are the nodal coordinates of the element.

The shape functions will be expressed through the natural coordinate system r, s and t.

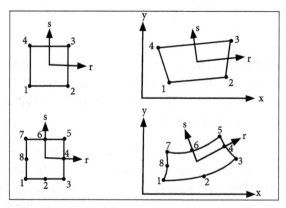

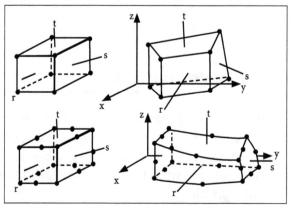

Isoparametric elements -coordinate transformation.

The shape functions $[N']$ used in the above transformation thus help us to define the geometry of the element in the Cartesian coordinate system. If these shape functions $[N']$ are the same as the shape functions $[N]$ used to represent the variation of displacement in the element, these elements are called Isoparametric elements.

$$\{x\} = [N]\{x_n\}$$

And in cases where the geometry of the element is defined by shape functions of order higher than that for representing the variation of displacements, the elements are called 'super parametric'. Similarly if more nodes are used to define displacement compared to the nodes used to represent the geometry of the elements, then they would be referred to as 'sub parametric' elements.

Sub Parametric and Super Parametric Elements

Generally it is very difficult to represent the curved boundaries by straight edge elements. A large number of elements may be used to obtain reasonable resemblance between original body and the assemblage. In order to overcome this drawback, Isoparametric elements are used.

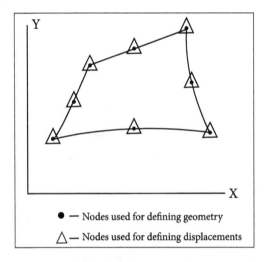

If the number of nodes used for defining the geometry is same as number of nodes used for defining the displacements, then it is known as Isoparametric element.

Super Parametric Element

If the number of nodes used for defining the geometry is more than number of nodes used for defining the displacements, then it is known as super parametric element.

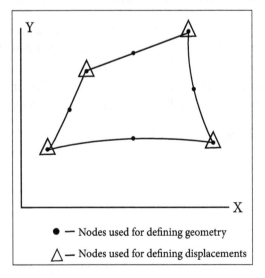

Sub Parametric Element

If the number of nodes used for defining the geometry is less than number of nodes used for defining the displacements, then it is known as sub parametric element.

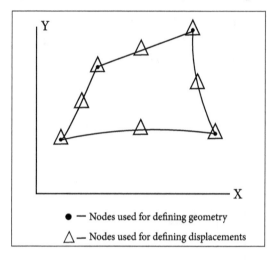

● — Nodes used for defining geometry

△ — Nodes used for defining displacements

Shape Functions For Isoparametric Elements

Isoparametric Elements are creating of the shape functions is normally the first and most common and important, steps in developing finite element equations for any type of element. In determining the shape functions Ni (i = 1, 2, 3) for the triangular element, followed by starting with an assumption of the displacements using polynomial basis functions with unknown constants.

These unknown constants are then found using the nodal displacements at the nodes of the element. This standard procedure works in principle for the development of any type of element, but might not be the most convenient method. We demonstrate here another slightly different approach for constructing shape functions.

We start with an assumption of shape functions directly using polynomial basis functions with unknown constants. These unknown constants are then determined using the property of the shape functions. The only difference here is that we assume directly the shape function instead of the displacements. For a linear triangular element, we assume that the shape functions are linear functions of x and y.

They should, therefore, have the form of:

$$N_1 = a_1 + b_1 x + c_1 y \qquad \qquad ...(1)$$

$$N_2 = a_2 + b_2 x + c_2 y \qquad \qquad ...(2)$$

$$N_3 = a_3 + b_3 x + c_3 y \qquad \qquad ...(3)$$

Where,

a_i, b_i and C_i (i = 1, 2, 3) are constants to be determined. Equation (1) can be written in a concise form:

$$N_i = a_i + b_i x + c_i y, \qquad i = 1, 2, 3 \qquad \qquad ...(4)$$

We write the shape functions in the following matrix form:

$$N_i = \underbrace{\{1 \quad x \quad y\}}_{p} \underbrace{\begin{Bmatrix} a_i \\ b_i \\ c_i \end{Bmatrix}}_{\alpha} = p^T \alpha \qquad \qquad ...(5)$$

Where,

      a - The vector of the 3 unknown constants.

      P - The vector of polynomial basis functions (or monomials).

Using above equations, the moment matrix P corresponding to basis p can be given by:

$$P = \begin{bmatrix} 1 & x_1 & y_1 \\ 1 & x_2 & y_2 \\ 1 & x_3 & y_3 \end{bmatrix} \qquad \qquad ...(6)$$

Note that the above equation is written for the shape functions and not for the displacements. For this particular problem, we use up to the first order of polynomial basis. Depending upon the problem, we could use higher order of polynomial basis functions.

The complete order of polynomial basis functions in two-dimensional space up to the ith order can be given by using the so-called Pascal triangle.

The number of terms used in p depends upon the number of nodes the 2D elements. We generally try to use terms of lowest orders to create the basis as complete as possible in order. It is also possible to choose specific terms of higher orders for different types of elements. For a triangular element there are 3 nodes and therefore the lowest terms with complete first order are used, as shown in equation (5).

The assumption of equation (1) implies that the displacement is assumed to vary linearly in the element. In equation (4) there is a total of 9 constants to be determined.

Therefore, we can expect that the complete linear basis functions used in equation (5) guarantee that the shape functions to be constructed satisfy the sufficient requirements for FEM shape functions. What we required to do now is simply impose the delta function property on the assumed shape functions to determine the unknown constants a_i, b_i and C_i.

The delta functions property states that the shape function should be a unit at its home node and zero at its remote nodes. For a two-dimensional problem, it can be expressed as:

$$N_i\left(x_j, y_j\right) = \begin{cases} 0 & \text{for } i = j \\ 0 & \text{for } i \neq j \end{cases} \quad\quad ...(7)$$

For a triangular element, this condition can be expressed explicitly for all 3 shape functions in the following equations. For shape function Ni, we have:

$$N_1\left(x_1, y_1\right) = 1$$

$$N_1\left(x_2, y_2\right) = 0$$

$$N_1\left(x_3, y_3\right) \quad 0 \quad\quad ....(8)$$

This is because node 1 at $(x_1, y_0$ is the home node of N1 and nodes 2 at (x_2, y_2) and 3 at (x_3, y_3) are the remote nodes of N_1. Using equations (1) and (9), we have:

$$N_1\left(x_1, y_1\right) = a_1 + b_1 x_1 + c_1 y_1 = 1$$

$$N_1\left(x_2, y_2\right) = a_1 + b_1 x_2 + c_1 y_2 = 0$$

$$N_1\left(x_3, y_3\right) = a_1 + b_1 x_3 + c_1 y_3 = 0 \quad\quad ...(9)$$

Solving expressions for a1, b1 and C1, we obtain:

$$a_1 = \frac{x_2 y_3 - x_3 y_2}{2 Ae}, \, b_1 = \frac{y_2 - y_3}{2 A_e}, \, c_1 = \frac{x_3 - x_2}{2 A_e} \quad\quad ...(10)$$

Where, Ae is the area of the triangular element that can be determined using the determinant of the moment matrix.

$$A_e = \frac{1}{2}|p| = \frac{1}{2}\begin{vmatrix} 1 & x_1 & y_1 \\ 1 & x_2 & y_2 \\ 1 & x_3 & y_3 \end{vmatrix} = \frac{1}{2}\left[(x_2 y_3 - x_3 y_2) + (y_2 - y_3)x_1 + (x_3 - x_2)y_1\right]...(11)$$

Note here that as long as the area of the triangular element is nonzero or as long as the 3 nodes are not on the same line, the moment matrix P will be of full rank. While substitution made over the expressions, we get:

$$N_1 = \frac{1}{2A_e}\left[(x_2 y_3 - x_3 y_2) + (y_2 - y_3)x + (x_3 - x_2)y\right]$$

Which can be re-written as:

$$N_1 = \frac{1}{2A_e}\left[(y_2 - y_3)(x - x_2) + (x_3 - x_2)(y - y_2)\right]$$

This equation clearly shows that N1 is a plane in the space of (x, y, N) that passes through the line of 2-3 and vanishes at nodes 2 at (x2, y2) and 3 at(x3, y3). This plane also passes the point of (x1, y1, 1) in space that guarantees the unity of the shape function at the home node.

Since the shape function varies linearly within the element, N1 can then be simply plotted as in below the figure. Making use of these features of N1, we can immediately write out the other 2 shape functions for nodes 2 and 3. For node 2, the conditions are:

$$N_2\left(x_1, y_1\right) = 0$$

$$N_2\left(x_2, y_2\right) = 1$$

$$N_2\left(x_3, y_3\right) = 0$$

And the shape function N2 should pass through the line 3-1, which gives:

$$N_2 = \frac{1}{2A_e}\left[(x_3 y_1 - x_1 y_3) + (y_3 - y_1)x + (x_1 - x_3)y\right]$$

$$= \frac{1}{2A_e}\left[(y_3 - y_1)(x - x_3) + (x_1 - x_3)(y - y_3)\right]$$

Which is plotted in below figure.

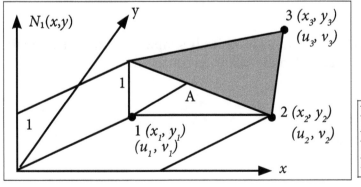

(a) Shape function N_1

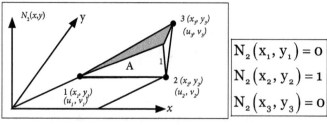

$$N_2(x_1, y_1) = 0$$
$$N_2(x_2, y_2) = 1$$
$$N_2(x_3, y_3) = 0$$

(b) Shape function N_2

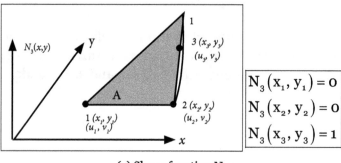

$$N_3(x_1, y_1) = 0$$
$$N_3(x_2, y_2) = 0$$
$$N_3(x_3, y_3) = 1$$

(c) Shape function N_3

For node 3, the conditions are:

$$N_3(x_1, y_1) = 0$$

$$N_3(x_2, y_2) = 0$$

$$N_3(x_3, y_3) = 1$$

And the shape function N3 should pass through the line 1-2 and given by:

$$N_3 = \frac{1}{2A_e}\left[(x_1 y_2 - x_1 y_1) + (y_1 - y_2)x + (x_2 - x_1)y\right]$$

$$= \frac{1}{2A_e}\left[(y_1 - y_2)(x - x_1) + (x_2 - x_1)(y - y_1)\right]$$

The process of finding these constants is basically simple, algebraic manipulation. The shape functions are summarized in the following concise form:

$$N_i = a_i + b_i x + c_i\, y$$

$$a_i = \frac{1}{2A_e}\left(x_j y_k - x_k y_j\right)$$

$$b_i = \frac{1}{2A_e}\left(y_j - y_k\right)$$

$$c_i = \frac{1}{2A_e}\left(x_k - x_j\right)$$

Where, the subscript i varies from 1 to 3 and j and k are calculated by the cyclic permutation in the order of i, j, k. For example, if i = 1, then j = 2, k = 3. When i = 2, then j = 3, k = 1.

One Dimensional

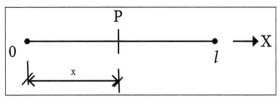

Cartesian coordinate system.

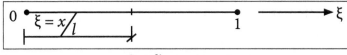

Coordinate system.

Shape functions for a two-node (linear) one dimensional element are given by:

$$N_1 = \frac{1}{2}(1-\xi), \qquad N_2 = \frac{1}{2}(1+\xi)$$

Where, ξ is the local coordinate defined as $\xi 2(x - x_c)/1$ in which l is length of the element and xc represents its midpoint. Here, nodes 1 and 2 correspond to $\xi = -1$ and $\xi = +1$ respectively. Similarly, shape functions for a quadratic (three-node) element are given by:

$$N_1 = \frac{1}{2}\xi(\xi-1), \ N_2 = \left(1-\xi^2\right) \quad N_3 = \frac{1}{2}\xi(1+\xi)$$

Where, nodes 1, 2 and 3 correspond to $\xi = -1$ and $\xi = +1$ respectively.

Two Dimensions

In two-dimensions, one can rectangular or triangular elements. In both case, shape functions are expressed in corresponding natural coordinates. These shape functions can also be used to define the physical coordinates of any point as follows:

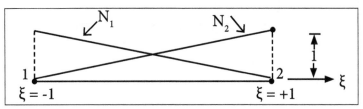

Shape function two-order elements.

$$x(\xi,\eta)=\sum_{\alpha=1}^{n}N_{\alpha}(\xi,\eta)x_{\alpha}, \ y(\xi,\eta)=\sum_{\alpha=1}^{n}N_{\alpha}(\xi,\eta)y_{\alpha}$$

Rectangular Elements

Shape functions for rectangular elements are expressed in terms of natural coordinates ξ and η defined as:

$$\xi=\frac{2(x-x_c)}{l_x}, \qquad \eta=\frac{2(y-y_c)}{l_y}$$

Where, (xc, yc) is the centroid of the element and lx, ly represent its extents in x and y-direction. Two-dimensional bilinear rectangular elements are given by:

$$N_{\alpha}=\frac{1}{4}(1+\xi\xi_{\alpha})(1+\eta\eta_{\alpha}), \qquad \alpha=1,..,4$$

Where, ξ and η are natural coordinates and coordinates of the four nodes in natural coordinates are (-1,-1), (1,-1), (1, 1) and (-1, 1).

Triangular Elements

Shape functions for triangular elements are expressed in terms of area coordinate ξ_1, ξ_2 and ξ_3 defined as:

$$\xi_1=\frac{a_1+b_1x+c_1y}{2\Delta}, \xi_2=\frac{a_2+b_2x+c_2y}{2\Delta}, \xi_3=1-(\xi_1+\xi_2)$$

Where, Δ is the area of the triangle with vertices $(x_1,y_1),(x_2,y_2),(x_3,y_3)$ and coefficients a_k, b_k and c_k are given by:

$$a_1=x_2y_3-x_3y_2, \ b_1=y_2-y_3, c_1=x_3-x_2$$

$$a_2=x_2y_3-x_3y_2, b_2=y_3-y_1, c_2=x_1-x_3$$

In terms of area coordinates, shape functions for a linear triangular element are defined as:

$$N_1=\xi_1, \ N_2=\xi_2, \ N_3=\xi_3$$

Shape functions for a quadratic triangular element are given by:

i) Corner nodes (numbered as 1, 2, 3)

$$N_1=\xi_1(2\xi_1-1), \ N_2=\xi_2(2\xi_2-1), \ N_3=\xi_3(2\xi_3-1)$$

ii) Mid- side nodes (numbered as 4, 5, 6)

$$N_4 = 4\xi_1\xi_2, \; N_5 = 4\xi_2\xi_3, \; N_6 = 4\xi_3\xi_1$$

3.3.1 Stiffness Matrix of Isoparametric Elements

The term Isoparametric is derived from the use of the same shape functions (or inter-polation functions) [N] to define the element's geometric shape as are used to define the displacements within the element.

Thus, when the shape function is u = a1+ a2s for the displacement, we use x=a1 + a2s for the description of the nodal coordinate of a point on the bar element and hence, the physical shape of the element.

Isoparametric element equations are formulated using a natural (or intrinsic) co-ordinate system 's' that is defined by element geometry and not by the element orientation in the global coordinate system.

In other words, axial coordinate 's' is attached to the bar and remains directed along the axial length of the bar, regardless of how the bar is oriented in space. There is a relationship (called a transformation mapping) between the natural coordinate system 's' and the global coordinate system 'x' for each element of a specific structure and this relationship must be used in the element equation formulations.

We will now develop the Isoparametric formulation of the stiffness matrix of a simple linear bar element [with two nodes as shown in the figure 1(a)].

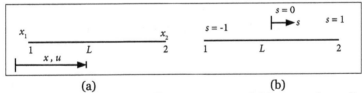

(a)   (b)

(1) Linear bar element in (a) a global coordinate system x and (b) a natural coordinate system s.

Step 1: Select Element Type

First, the natural coordinates is attached to the element, with the origin located at the center of the element, as shown in figure 1(b). We consider the bar element to have two degrees of freedom axial displacements u1 and u2 at each node associated with the global x axis.

For the special case when the s and x axes are parallel to each other, the s and x coordi-nates can be related by:

$$x = x_c + \frac{L}{2}s \qquad \qquad \qquad ...(1a)$$

Where,

x_c - Global coordinate of the element centroid.

Using the global coordinates x_1 and x_2 in equation (1a) with $x_c = (x_1 + x_2)/2$, we can express the natural coordinate s in terms of the global coordinates as:

$$s = \left[x - (x_1 + x_2)/2\right]\left(2/(x_2 - x_1)\right) \qquad \text{...(1b)}$$

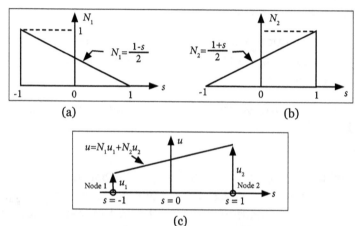

(2) Shape function variations with natural coordinates: (a) shape function N1, (b) shape function N2 and (c) linear displacement field u plotted over element length.

We begin by relating the natural coordinate to the global coordinate by:

$$x = a_1 + a_2 s \qquad \text{....(2)}$$

Where, we note that 's' is such that $- \leq s \leq 1$. Solving for the ai's in terms of x1 and x2, we obtain:

$$x = \frac{1}{2}\left[(1-s)x_1 + (1+s)x_2\right] \qquad \text{...(3)}$$

or, in matrix form, we can express equation (3) as:

$$\{x\} = \begin{bmatrix} N_1 & N_2 \end{bmatrix} \begin{Bmatrix} x_1 \\ x_2 \end{Bmatrix} \qquad \text{...(4)}$$

where, the shape functions in equation (4) are:

$$N_1 = \frac{1-s}{2} \qquad N_2 = \frac{1+s}{2} \qquad \text{...(5)}$$

The linear shape functions in equations (5) map the 's' coordinate of any point in the element to the 'x' coordinate when used in equation (3). For instance, when we substitute s=-1 into equation (3), we obtain x=x_1.

These shape functions are shown in the figure(2).Hence,N_1 represents the physical shape of the coordinate x when plotted over the length of the element for $x_1=1$ and $x_2=0$ and N_2 represents the coordinate 'x' when plotted over the length of the element for $x_2=1$ and $x_1=0$. Again, we must have $N_1+N_2=1$.

These shape functions must also be continuous throughout the element domain and have finite first derivatives within the element.

Step 2: Select a Displacement Function

The displacement function within the bar is now defined by the same shape functions, equations (5), as are used to define the element shape, that are given by:

$$\{u\}=\begin{bmatrix} N_1 & N_2 \end{bmatrix}\begin{Bmatrix} u_1 \\ u_2 \end{Bmatrix} \qquad ...(6)$$

When a particular coordinates of the point of interest is substituted into [N], equation (6) yields the displacement of a point on the bar in terms of the nodal degrees of freedom u1 and u2 as shown in the figure 2(c). Since u and x are defined by the same shape functions at the same nodes, comparing equations (4) and (6), the element is called Isoparametric.

Step 3: Define the Strain/Displacement and Stress/Strain Relationships

We now want to formulate element matrix [B] to evaluate [k]. We use the Isoparametric formulation to illustrate its manipulations. For a simple bar element, however, for higher-order elements, the advantage will become clear because, relatively simple computer program formulations will result.

To construct the element stiffness matrix, we must determine the strain, which is defined in terms of the derivative of the displacement with respect to x. The displacement u, however, is now a function of 's' as given by equation (6). Therefore, we must apply the chain rule of differentiation to the function u as follows:

$$\frac{du}{ds}=\frac{du}{dx}\frac{dx}{ds} \qquad ...(7)$$

We can evaluate (du/ds) and (dx/ds) using equations (6) and (3). We seek $(du/dx)=\varepsilon_x$. Therefore, we solve equation (7) for (du/dx) as:

$$\frac{du}{dx}=\frac{\left(\dfrac{du}{ds}\right)}{\left(\dfrac{dx}{ds}\right)} \qquad ...(8)$$

Using equation (6) for u, we obtain:

$$\frac{du}{ds} = \frac{u_2 - u_1}{2} \qquad ...(9a)$$

And using equation (3) for x, we have:

$$\frac{dx}{ds} = \frac{x_2 - x_1}{2} = \frac{L}{2} \qquad ...(9b)$$

Because, $x_2 - x_1 = L$.

Using equations (9a) and (9b) in equation (8), we obtain:

$$\{\varepsilon_x\} = \begin{bmatrix} -\dfrac{1}{L} & \dfrac{1}{L} \end{bmatrix} \begin{Bmatrix} u_1 \\ u_2 \end{Bmatrix} \qquad ...(10)$$

Since $\{\varepsilon\} = [B] \ \{d\}$, the strain/displacement matrix [B] is then given in equation (10) as:

$$[B] = \begin{bmatrix} -\dfrac{1}{L} & \dfrac{1}{L} \end{bmatrix} \qquad ...(11)$$

We recall that use of linear shape functions results in a constant B matrix and hence, in a constant strain within the element. For higher-order elements, such as the quadratic bar with three nodes,[B] becomes a function of natural coordinates.

The stress matrix is again given by Hooke's law as:

$$\underline{\sigma} = E\underline{\varepsilon} = E\,\underline{B}\,\underline{d}$$

Step 4: Derive the Element Stiffness Matrix and Equations

The stiffness matrix is given by:

$$[k] = \int_0^L [B]^T [D][B] A \, dx \qquad ...(12)$$

However, in general, we must transform the coordinate x to s because [B] is, in general, a function of s. This general type of transformation is given by:

$$\int_0^L f(x)\,dx = \int_{-1}^1 f(s)|\underline{J}|\,ds \qquad ...(13)$$

Where $\underline{J}$ is called the Jacobian. In the one-dimensional case, we have $|\underline{J}| = \underline{J}$. For the simple bar element, from equation (9b), we have:

$$|\underline{J}| = \frac{dx}{ds} = \frac{L}{2} \qquad ...(14)$$

Observe that in equation (14), the Jacobian relates an element length in the global-co-ordinate system to an element length in the natural-coordinate system. In general, $|J|$ a function of s and depends on the numerical values of the nodal coordinates.

Using equations (13) and (14) in equation (12), we obtain the stiffness matrix in natural coordinates as:

$$[k] = \frac{L}{2} \int_{-1}^{L} [B]^T E[B] A \, ds \qquad \qquad ...(15)$$

where, for the one-dimensional case, we have used the modulus of elasticity E=[D] in equation (15). Substituting equation (11) in equation (15) and performing the simple integration, we obtain:

$$[k] = \frac{AE}{L} \begin{bmatrix} 1 & -1 \\ -1 & 1 \end{bmatrix} \qquad \qquad ...(16)$$

For higher-order one-dimensional elements, the integration in closed form becomes difficult if not impossible. Even the simple rectangular element stiffness matrix is difficult to evaluate in closed form.

Body Forces

We will now determine the body-force matrix using the natural coordinate systems. The body-force matrix is given by:

$$\{\hat{f}_b\} = \iiint_V [N]^T \{\hat{X}_b\} dV \qquad \qquad ...(17)$$

Letting dV = Adx, we have:

$$\{\hat{f}_b\} = A \int_0^L [N]^T \{\hat{X}_b\} dx \qquad \qquad ...(18)$$

Substituting equations (5) for N_1 and N_2 into [N] and noting that by equation (9b), dx = (L/2) ds, we obtain as:

$$\{\hat{f}_b\} = A \int_{-1}^{1} \begin{Bmatrix} \dfrac{1-s}{2} \\ \dfrac{1+s}{2} \end{Bmatrix} \{\hat{X}_b\} \frac{L}{2} ds \qquad \qquad ...(19)$$

On integrating equation (19), we obtain:

$$\{\hat{f}_b\} = \frac{A L \hat{X}_b}{2} \begin{Bmatrix} 1 \\ 1 \end{Bmatrix} \qquad \qquad ...(20)$$

The physical interpretation of the results for $\{\hat{f}_b\}$ is that since AL represents the volume of the element and $\hat{X}_b$ the body force per unit volume, then $A L \hat{X}_b$ is the total body force acting on the element. The factor 1/2 indicates that this body force is equally distributed to the two nodes of the element.

Surface Forces

Surface forces can be taken as:

$$\{\hat{f}_s\} = \iint_S [N_s]^T \{\hat{T}_x\} dS \qquad \qquad ...(21)$$

Assuming the cross section is constant and the traction is uniform over the perimeter and along the length of the element, we obtain:

$$\{\hat{f}_s\} = \int_0^L [N_s]^T \{\hat{T}_x\} dx \qquad ... \ 22 \qquad ...(22)$$

Where we now assume is in units of force per unit length. Using the shape functions N_1 and N_2 from equation (5) in equation (22), we obtain:

$$\{\hat{f}_s\} = \int_{-1}^1 \begin{Bmatrix} \dfrac{1-s}{2} \\ \dfrac{1+s}{2} \end{Bmatrix} \{\hat{T}_x\} \dfrac{L}{2} ds \qquad ...(23)$$

On integrating equation (23), we obtain:

$$\{\hat{f}_s\} = \{\hat{T}_x\} \dfrac{L}{2} \begin{Bmatrix} 1 \\ 1 \end{Bmatrix} \qquad ...(24)$$

The physical interpretation of equation (24) is that since $\hat{T}_x$ is in force-per-unit-length units, $\hat{T}_x$ is now the total force. Then 1/2 indicates that the uniform surface traction is equally distributed to the two nodes of the element.

Note that if $\hat{T}_x$ were a function of x (or s), then the amounts of force allocated to each node would generally not be equal.

3.4   Three Dimensional Elements

As in the two-dimensional case, there are two main families of three-dimensional elements. One is based on extension of triangular elements to tetrahedrons and the other on extension of rectangular elements to rectangular parallelopipeds.

The algebraically cumbersome technique for deriving interpolation functions in global Cartesian coordinates has been illustrated for two-dimensional elements. Those developments are not repeated here for three-dimensional elements, the procedures are algebraically identical but even more complex. Instead, we utilize only the more amenable approach of using natural coordinates to develop the interpolation functions for the two basic elements of the tetrahedral and brick families.

Four-Node Tetrahedral Element

A four-node tetrahedral element is depicted in the figure (a), in relation to a global Cartesian coordinate system. The nodes are numbered 1-4 per the convention that node 1 can be selected arbitrarily and nodes 2-4 are then specified in a counterclockwise direction from node 1.

This convention is the same as used by most commercial finite element analysis software and is very important in assuring geometrically correct tetrahedrons. On the other hand, tetrahedral element definition for finite element models is so complex that it is almost always accomplished by auto meshing capabilities of specific software packages.

In a manner analogous to use of area coordinates, we now introduce the concept of volume coordinates using in figure (b). Point P(x, y. z) is an arbitrary point in the tetrahedron defined by the four nodes. As indicated by the dashed lines, point P and the four nodes define four other tetrahedral having volumes:

$$v_1 = vol(P234) \quad v_2 = vol(P134)$$
$$v_3 = vol(P124) \quad v_4 = vol(P123)$$

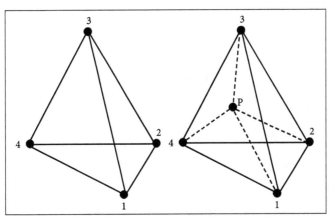

(a) A four node tetrahedral elements (b) An arbitrary interest point.

The volume coordinate are defined as:

$$L_1 = \frac{V_1}{V}$$

$$L_2 = \frac{V_2}{V}$$

$$L_3 = \frac{V_3}{V}$$

$$L_4 = \frac{V_4}{V}$$

Where,

V- total volume of the elements given by:

$$V = \frac{1}{6} \begin{vmatrix} 1 & x_1 & y_1 & z_1 \\ 1 & x_2 & y_2 & z_2 \\ 1 & x_3 & y_3 & z_3 \\ 1 & x_4 & y_4 & z_4 \end{vmatrix}$$

As with area coordinates, the volume coordinates are not independent, since:

$$v_1 + v_2 + v_3 + v_4 = v$$

Now, let us examine the variation of the volume coordinates through the element. If, for example, point P corresponds to node 1, we find $V_1 = V$, $V_2 = V_3 = V_4 = 0$. Consequently $L_1 = 1$, $L_2 = L_3 = L_4 = 0$ at node 1. As P moves away from node 1, V1 decreases linearly, since the volume of a tetrahedron is directly proportional to its height (the perpendicular distance from P to the plane defined by nodes 2, 3 and 4) and the area of its base (the triangle formed by nodes 2, 3 and 4).

On any plane parallel to the base triangle of nodes 2, 3, 4, the value of L1 is constant. A particular importance is that, if P lies in the plane of nodes 2, 3, 4, the value of L_1 is zero. Identical observations apply to volume coordinates L_2, L_3 and L_4. So the volume coordinates satisfy all required nodal conditions for interpolation functions and we can express the field variable as:

$$\phi(x,y,z) = L_1\phi_1 + L_2\phi_2 + L_3\phi_3 + L_4\phi_4 \qquad \qquad ...(1)$$

Explicit representation of the interpolation functions (i.e., the volume co-ordinates) in terms of global coordinates is, as stated, algebraically complex but straightforward. Fortunately, such explicit representation is not generally required; as element formulation can be accomplished using volume coordinates only.

As with area coordinates, integration of functions of volume coordinates (required in developing element characteristic matrices and load vectors) is relatively simple. Integrals of the form:

$$\iiint_{V} L_1^a L_2^b L_3^c L_4^d \, dV$$

Where, a, b, c, d are positive integers and V is total element volume, appear in element formulation for various physical problems. As with area coordinates. Integration in volume coordinates is straightforward and we have the integration formula:

$$\iiint_{V} L_1^a L_2^b L_3^c L_4^d \, dV = \frac{a!b!c!d!}{(a+b+c+d+3)!} \, (6V)$$

The is three-dimensional analogy to equation:

As another analogy with the two-dimensional triangular elements, the tetrahedral element is most useful in modeling irregular geometries. However, the tetrahedral element is not particularly amenable to use in conjunction with other element types, strictly as a result of the nodal configurations.

As a final comment on the four-node tetrahedral element, we note that the field variable representation, as given by equation (1), is a linear function of the Cartesian coordinates. Therefore, all the first partial derivatives of the field variable are constant. In structural applications, the tetrahedral element is a constant strain element; in general, the element exhibits constant gradients of the field variable in the coordinate directions.

On other elements of the tetrahedral family are depicted in the figure(c). The interpolation functions for the depicted elements are readily written in volume coordinates, as for higher-order two-dimensional triangular elements. Note particularly that the second element of the family has 10 nodes and a cubic form for the field variable and interpolation functions. A quadratic tetrahedral element cannot be constructed to exhibit geometric isotropy even if internal nodes are included.

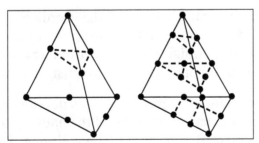

Higher order tetrahedral elements (c) 10 node (d) 20 node.

Eight-Node Brick Element

The so-called eight node brick element is shown in the figure (a) in reference to a global

Cartesian coordinate system. Here, we utilize the natural coordinate's r, s, t of in figure (b) defined as:

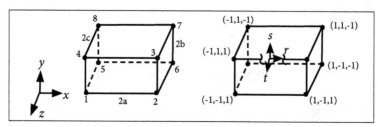

Eight-node brick elements: (a) global Cartesian coordinate. (b) Natural coordinate with an origin at the centroid.

$$r = \frac{x - \overline{x}}{a}$$

$$s = \frac{y - \overline{y}}{b}$$

$$t = \frac{z - \overline{z}}{c}$$

Where, 2_a, 2_b, 2_c are the dimensions of the element in the x, y, z coordinates respectively and the coordinates of the element centroid are:

$$\overline{x} = \frac{x_2 - x_1}{2}$$

$$\overline{y} \quad \frac{y_3 \quad y_2}{}$$

$$\overline{z} = \frac{z_5 - z_1}{2}$$

The natural coordinates are defined such that the coordinate values vary between -1 and +1 over the domain of the element. As with the plane rectangular element, the natural coordinates provide for a straightforward development of the interpolation functions by using the appropriate monomial terms to satisfy nodal conditions.

As we illustrated the procedure in several previous developments, Instead, we simply write the interpolation functions in terms of the natural coordinates and request that the reader verify satisfaction of all nodal conditions. The interpolation functions are:

$$N_1 = \frac{1}{8}(1-r)(1-s)(1+t)$$

$$N_2 = \frac{1}{8}(1+r)(1-s)(1+t)$$

$$N_3 = \frac{1}{8}(1+r)(1+s)(1+t)$$

$$N_4 = \frac{1}{8}(1-r)(1+s)(1+t)$$

$$N_5 = \frac{1}{8}(1-r)(1-s)(1-t)$$

$$N_6 = \frac{1}{8}(1+r)(1-s)(1-t)$$

$$N_7 = \frac{1}{8}(1+r)(1+s)(1-t)$$

$$N_8 = \frac{1}{8}(1-r)(1+s)(1-t)$$

And the field variable is described as:

$$\phi(x,y,z) = \sum_{i=1}^{8} N_i(r,s,t)\phi_i \qquad \qquad ...(2)$$

If equation (2), is expressed in terms of the global Cartesian coordinates, it is found to be of the form:

$$\phi(x,y,z) = a_0 + a_1 x + a_2 y + a_3 z + a_4 xy + a_5 xz + a_6 yz + a_7 xyz \qquad ...(3)$$

Showing that the field variable is expressed as an incomplete, symmetric polynomial. Geometric isotropy is therefore assured. The compatibility requirement is satisfied, as is the completeness condition. Recall that completeness requires that the first partial derivatives must be capable of assuming constant values (for C° problems). If, for example, we take the first partial derivative of Equation (3), with respect to x, we obtain:

$$\frac{\partial \phi}{\partial x} = a_1 + a_4 y + a_5 z + a_7 yz \qquad \qquad ...(4)$$

Which certainly does not appear to be constant at first glance. However, if we apply the derivative operation to equation (2), while noting that:

$$\frac{\partial \phi}{\partial x} = \frac{1}{a}\frac{\partial \phi}{\partial r} \qquad \qquad ...(5)$$

The result is given by:

$$\frac{\partial \phi}{\partial x} = \frac{1}{8a}(1-s)(1+t)(\phi_2 - \phi_1) + \frac{1}{8a}(1+s)(1+t)(\phi_3 - \phi_4)$$

$$+ \frac{1}{8a}(1-s)(1-t)(\phi_6 - \phi_5) + \frac{1}{8a}(1+s)(1-t)(\phi_7 - \phi_8) \qquad ...(6)$$

Referring to figure (a), observe that, if the gradient of the field variable in the x direction is constant. $\partial\phi/\partial x = C$. the nodal values are related by:

$$\phi_2 = \phi_1 + \frac{\partial \phi}{\partial x} dx = \phi_1 + C(2a)$$

$$\phi_3 = \phi_4 + \frac{\partial \phi}{\partial x} dx = \phi_4 + C(2a)$$

$$\phi_6 = \phi_5 + \frac{\partial \phi}{\partial x} dx = \phi_5 + C(2a)$$

$$\phi_7 = \phi_8 + \frac{\partial \phi}{\partial x} dx = \phi_8 + C(2a)$$

Substituting these relations into equation (6), we find:

Which, on expansion and simplification, results in given by:

$$\frac{\partial \phi}{\partial x} \equiv C$$

Observing that this result is valid at any point r, s, o within the element, it follows that the specified interpolation functions indeed allow for a constant gradient in the x direction. Following similar procedures shows that the other partial derivatives also satisfy the completeness condition.

3.4.1 Isoparametric Formulations

The finite element method is a powerful technique for analyzing engineering problems involving complex, irregular geometries. Consider the plane area shown in the figure (a), which is to be discretized via a mesh of finite elements.

A possible mesh using triangular elements is shown in the figure (b). Note that the outermost "row" of elements provides a chordal approximation to the circular boundary and as the size of the elements is decreased, the approximation becomes increasingly closer to the actual geometry.

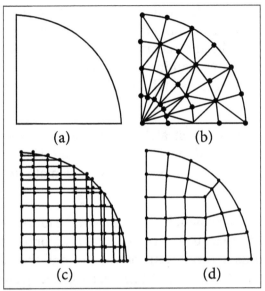

a) A domain to be modeled. (b) Triangular elements. (c) Rectangular elements. (d) Rectangular and quadrilateral elements.

However, also note that the elements in the inner "rows" become increasingly slender (i.e., the height to base ratio is large). In general, the ratio of the largest characteristic dimension of an element to the smallest characteristic dimension is known as the aspect ratio. Large aspect ratios increase the inaccuracy of the finite element representation and have a detrimental effect on convergence of finite element solutions.

An aspect ratio of 1 is ideal but cannot always be maintained shown in the figure(b), to maintain a reasonable aspect ratio for the inner elements, it would be necessary to reduce the height of each row of elements as the center of the sector is approached.

This observation is also in keeping with the convergence requirements of the h-refinement method. Although the triangular element can be used to closely approximate a curved boundary, other considerations dictate a relatively large number of elements and associated computation time.

If we consider rectangular elements as in the figure(c) (an intentionally crude mesh for illustrative purposes), the problems are apparent. Unless the elements are very small, the area of the domain excluded from the model (the shaded area in the figure) may be significant. For the case depicted, a large number of very small square elements best approximates the geometry.

The shaded areas of the figure (c) could be modeled by three-node triangular elements. Such combination of element types may not be the best in terms of solution accuracy since the rectangular element and the triangular element have, by necessity, different order polynomial representations of the field variable. The field variable is continuous across such element boundaries; this is guaranteed by the finite element formulation.

However, conditions on derivatives of the field variable for the two element types are quite different. On a curved boundary such as that shown, the triangular element used to fill the "gaps" left by the rectangular elements may also have adverse aspect ratio characteristics. Now examine in the figure (d), which shows the same area meshed with rectangular elements and a new element applied near the periphery of the domain.

The new element has four nodes, straight sides, but is not rectangular. The new element is known as a general two-dimensional quadrilateral element and is seen to mesh ideally with the rectangular element as well as approximate the curved boundary, just like the triangular element.

The four-node quadrilateral element is derived from the four-node rectangular element (known as the parent element) element via a mapping process. The figure (e) shows the parent element and its natural (r, s) coordinates and the quadrilateral element in a global Cartesian coordinate system. The geometry of the quadrilateral element is described by:

$$x = \sum_{i=1}^{4} G_i(x,y) x_i \qquad \qquad ...(1)$$

$$y = \sum_{i=1}^{4} G_i(x,y) y_i \qquad \qquad ...(2)$$

Where, the Gi (x, y) can be considered as geometric interpolation functions and each such function is associated with a particular node of the quadrilateral element.

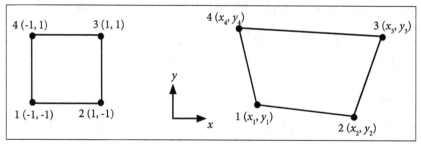

(e) Mapping of a parent element into an isobarometric element. A rectangle is shown for example.

Given the geometry and the form of equations (1) and (2), each function G, (x, y) must evaluate to unity at its associated node and to zero at each of the other three nodes.

These conditions are exactly the same as those imposed on the interpolation functions of the parent element. Consequently, the interpolation functions for the parent element can be used for the geometric functions, if we map the coordinates so that:

$$(r,s) = (-1,-1) \Rightarrow (x_1, y_1)$$

$$(r,s)=(1,-1) \Rightarrow (x_2, y_2)$$

$$(r,s)=(1, 1) \Rightarrow (x_3, y_3)$$

$$(r,s)=(-1, 1) \Rightarrow (x_4, y_4) \qquad \qquad ...(3)$$

Where, the symbol is read as "maps to" or "corresponds to." Note that the (r, s) coordinates used here are not the same as those defined by equations. In-stead, these are the actual rectangular coordinates of the 2 unit by 2 unit parent element.

Consequently, the geometric expressions become:

$$x = \sum_{i=1}^{4} N_i(r,s)x_i$$

$$y = \sum_{i=1}^{4} N_i(r,s)y_i$$

Clearly, we can also express the field variable variation in the quadrilateral element as:

$$\phi(x,y)=\phi(r,s)=\sum_{i=1}^{4} N_i(r,s)\phi_i$$

If the mapping of the equation (3) is used, since all required nodal conditions are satisfied. Since the same interpolation functions are used for both the field variable and description of element geometry, the procedure is known as Isoparametric (constant parameter) mapping. The element defined by such a procedure is known as an Isoparametric element.

3.5 Axisymmetric Elements

Axisymmetric element is quite useful when symmetry with respect to geometry and loading exists about an axis of the body being analyzed. Problems that involve soil masses subjected to circular footing loads or thick-walled pressure vessels can often be analyzed using the axisymmetric element.

We begin with the development of the stiffness matrix for the simplest axisymmetric element, the triangular torus, whose vertical cross section is a plane triangle.

We then present the longhand solution of a thick-walled pressure vessel to illustrate the use of the axisymmetric element equations. This is followed by a description of some typical large scale problems that have been modeled using the axisymmetric element.

Derivation of the Stiffness Matrix

Axisymmetric elements are triangular tori such that each element is symmetric with respect to geometry and loading about an axis such as the z axis in the figure (a).

Hence, the z axis is called the axis of symmetry or the axis of revolution. Each vertical cross section of the element is a plane triangle. The nodal points of an axisymmetric triangular element describe circumferential lines, as indicated in the figure (a).

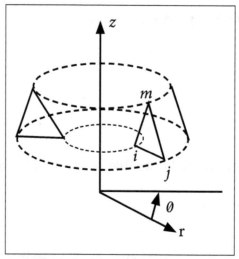

(a) Typical axisymmetric element ijm.

In plane stress problems, stresses exist only in the x-y plane. In axisymmetric problems, the radial displacements develop circumferential strains that induce stresses σr, σ_θ, σ_z and τ_{rz}, where r, θ and z indicate the radial, circumferential and longitudinal directions, respectively.

Triangular torus elements are often used to idealize the axisymmetric system because; they can be used to simulate complex surfaces and are simple to work with.

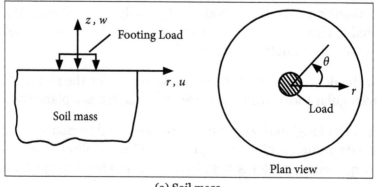

(a) Soil mass

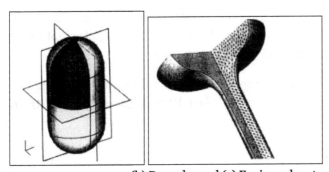

(b) Domed vessel (c) Engine valve stem.
(1)Examples of axisymmetric problems: (a) semi-infinite half-space (soil mass).
modeled by axisymmetric elements, (b) a domed pressure vessel and (c) an engine valve stem.

For instance, the axisymmetric problem of a semi-infinite half-space loaded by a circular area (circular footing) shown in the figure 1(a), the domed pressure vessel shown in the figure 1(b) and the engine valve stem shown in the figure 1(c) can be solved using the axisymmetric element.

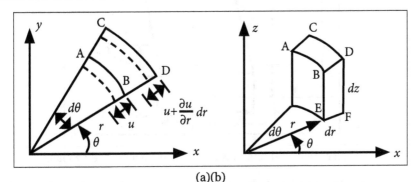

(a)(b)
(2) (a) Plane cross section (b) Axisymmetric element.

Because of symmetry about the z axis, the stresses are independent of the θ coordinate. Therefore, all derivatives with respect to θ vanish and the displacement component v (tangent to the θ direction), the shear strains $\gamma r\theta$ and $\gamma\theta z$ and the shear stresses $\tau r\theta$ and $\tau\theta z$ are all zero.

Figure (2) shows an axisymmetric ring element and its cross section to represent the general state of strain for an axisymmetric problem. It is most convenient to express the displacements of an element ABCD in the plane of a cross section in cylindrical coordinates.

We then let u and w denote the displacements in the radial and longitudinal directions, respectively. The side AB of the element is displaced an amount u and side CD is then displaced an amount $u +(\partial u/\partial r)$ in the radial direction. The normal strain in the radial direction is then given by:

$$\varepsilon_r = \frac{\partial u}{\partial r}$$ ...(1)

In general, the strain in the tangential direction depends on the tangential displacement v and on the radial displacement u. However, for axisymmetric deformation behavior, recall that the tangential displacement v is equal to zero.

Hence, the tangential strain is due only to the radial displacement. Having only radial displacement u, the new length of the are is (r+u) dθ and the tangential strain is then given by:

$$\varepsilon_\theta = \frac{(r+u)d\theta - r\,d\theta}{r\,d\theta} = \frac{u}{r} \qquad \qquad ...(2)$$

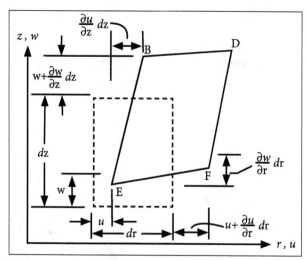

(b) Displacement and rotations of lines of element in the r-z plane.

Next, we consider the longitudinal element BDEF to obtain the longitudinal strain and the shear strain. In the figure (b), the element is shown to displace by amounts u and w in the radial and longitudinal directions at point E and to displace additional amounts $(\partial w/\partial z)\,\partial z$ along line BE and $(\partial u/\partial r)\,\partial r$ along line EF.

Furthermore, observing lines EF and BE, we see that point F moves upward an amount $(\partial w/\partial r)\partial r$ with respect to point E and point B moves to the right an amount $(\partial u/\partial z)\partial z$ with respect to point E.

Again, from the basic definitions of normal and shear strain, we have the longitudinal normal strain given by:

$$\varepsilon_z = \frac{\partial w}{\partial z} \qquad \qquad ...(3)$$

and the shear strain in the r-z plane given by:

$$\gamma_{rz} = \frac{\partial u}{\partial z} + \frac{\partial w}{\partial r} \qquad \qquad ...(4)$$

Summarizing the strain/displacement relationships of equations (1-4) in one equation for easier reference, we have:

$$\varepsilon_r = \frac{\partial u}{\partial r} \qquad \varepsilon_\theta = \frac{u}{r} \qquad \varepsilon_z = \frac{\partial w}{\partial z} \qquad \gamma_{rz} = \frac{\partial u}{\partial z} + \frac{\partial w}{\partial r} \quad ...(5)$$

The isotropic stress/strain relationship, obtained by simplifying the general stress/strain relationships is:

$$\left\{ \begin{array}{c} \sigma_r \\ \sigma_z \\ \sigma_\theta \\ \tau_{rz} \end{array} \right\} = \frac{E}{(1+v)(1-2v)} \begin{bmatrix} 1-v & v & v & 0 \\ v & 1-v & v & 0 \\ v & v & 1-v & 0 \\ 0 & 0 & 0 & \frac{1-2v}{2} \end{bmatrix} \left\{ \begin{array}{c} \varepsilon_r \\ \varepsilon_z \\ \varepsilon_\theta \\ \gamma_{rz} \end{array} \right\} \qquad ...(6)$$

Step 1: Select Element Type

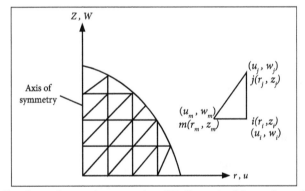

(a) Typical slice through an axisymmetric solid discretized into triangular elements.

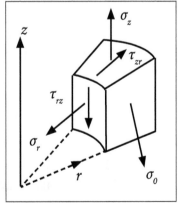

(b) Stresses in the axisymmetric problem.

(3)

An axisymmetric solid is shown discretized in the figure 3(a), along with a typical triangular element. The element has three nodes with two degrees of freedom per node (that is, ui, wi at node i). The stresses in the axisymmetric problem are shown in the figure3 (b).

Step 2 Select Displacement Functions

The element displacement functions are taken to be:

$$u(r,z) = a_1 + a_2 r + a_3 z$$

$$w(r,z) = a_4 + a_5 r + a_6 z \qquad \qquad ...(7)$$

So that we have the same linear displacement functions as used in the plane stress, constant-strain triangle. Again, the total number of ai' s (six) introduced in the displacement functions is the same as the total number of degrees of freedom for the element. The nodal displacements are:

$$\{d\} = \begin{Bmatrix} \underline{d}_i \\ \underline{d}_j \\ \underline{d}_m \end{Bmatrix} = \begin{Bmatrix} u_i \\ w_i \\ u_j \\ w_j \\ u_m \\ w_m \end{Bmatrix} \qquad \qquad ...(8)$$

And u evaluated at node i is:

$$u(r_i, z_i) = u_i = a_1 + a_2 r_i + a_3 z_i$$

Using equation (7), the general displacement function is then expressed in matrix form as:

$$\{\psi\} = \begin{Bmatrix} u \\ w \end{Bmatrix} = \begin{Bmatrix} a_1 + a_2 r + a_3 z \\ a_4 + a_5 r + a_6 z \end{Bmatrix} = \begin{bmatrix} 1 & r & z & 0 & 0 & 0 \\ 0 & 0 & 0 & 1 & r & z \end{bmatrix} \begin{Bmatrix} a_1 \\ a_2 \\ a_3 \\ a_4 \\ a_5 \\ a_6 \end{Bmatrix} \qquad ...(9)$$

Substituting the coordinates of the nodal points shown in the figure3 (a) into equation(9),we can solve the ai's. The resulting expressions are:

$$\begin{Bmatrix} a_1 \\ a_2 \\ a_3 \end{Bmatrix} = \begin{bmatrix} 1 & r_i & z_i \\ 1 & r_j & z_j \\ 1 & r_m & z_m \end{bmatrix}^{-1} \begin{Bmatrix} u_i \\ u_j \\ u_m \end{Bmatrix} \qquad \qquad ...(10)$$

And
$$
\begin{Bmatrix} a_4 \\ a_5 \\ a_6 \end{Bmatrix} = \begin{bmatrix} 1 & r_i & z_i \\ 1 & r_j & z_j \\ 1 & r_m & z_m \end{bmatrix}^{-1} \begin{Bmatrix} w_i \\ w_j \\ w_m \end{Bmatrix}
$$ ...(11)

Performing the inversion operations in equations (10) and (11), we have:

$$
\begin{Bmatrix} a_1 \\ a_2 \\ a_3 \end{Bmatrix} = \frac{1}{2A} \begin{bmatrix} \alpha_i & \alpha_j & \alpha_m \\ \beta_i & \beta_j & \beta_m \\ \gamma_i & \gamma_j & \gamma_m \end{bmatrix} \begin{Bmatrix} u_i \\ u_j \\ u_m \end{Bmatrix}
$$ ...(12)

And
$$
\begin{Bmatrix} a_4 \\ a_5 \\ a_6 \end{Bmatrix} = \frac{1}{2A} \begin{bmatrix} \alpha_i & \alpha_j & \alpha_m \\ \beta_i & \beta_j & \beta_m \\ \gamma_i & \gamma_j & \gamma_m \end{bmatrix} \begin{Bmatrix} w_i \\ w_j \\ w_m \end{Bmatrix}
$$ ...(13)

Where,

$$
\begin{array}{lll}
\alpha_i = r_j z_m - z_j r_m & \alpha_j = r_m z_i - z_m r_i & \alpha_m = r_i z_j - z_i r_j \\
\beta_i = z_j - z_m & \beta_j = z_m - z_i & \beta_m = z_i - z_j \\
\gamma_i = r_m - r_j & \gamma_j = r_i - r_m & \gamma_m = r_j - r_i
\end{array}
$$ (a)

We define the shape functions as:

$$
N_i = \frac{1}{2A} \left(\alpha_i + \beta_i r + \gamma_i z \right)
$$

$$
N_j = \frac{1}{2A} \left(\alpha_j + \beta_j r + \gamma_j z \right)
$$

$$
N_m = \frac{1}{2A} \left(\alpha_m + \beta_m r + \gamma_m z \right)
$$ ...(14)

Substituting equations (10) and (11) into equation (9), along with the shape function equations (14), we find that the general displacement function is:

$$
\{\psi\} = \begin{Bmatrix} u(r,z) \\ w(r,z) \end{Bmatrix} = \begin{bmatrix} N_i & 0 & N_j & 0 & N_m & 0 \\ 0 & N_i & 0 & N_j & 0 & N_m \end{bmatrix} \begin{Bmatrix} u_i \\ w_i \\ u_j \\ w_j \\ u_m \\ w_m \end{Bmatrix}
$$ ...(15)

Or $\{\Psi\} = [N] \{d\}$

Step 3: Define the Strain/Displacement and Stress/Strain Relationships

When we use equations (1-5) and (7), the strains become:

$$\{\varepsilon\} = \begin{Bmatrix} a_2 \\ a_6 \\ \dfrac{a_1}{r} + a_2 + \dfrac{a_3 z}{r} \\ a_3 + a_5 \end{Bmatrix} \qquad \text{...(16)}$$

Rewriting equation (16) with the a_i 's as a separate column matrix, we have:

$$\begin{Bmatrix} \varepsilon_r \\ \varepsilon_z \\ \varepsilon_\theta \\ \gamma_{rz} \end{Bmatrix} = \begin{bmatrix} 0 & 1 & 0 & 0 & 0 & 0 \\ 0 & 0 & 0 & 0 & 0 & 1 \\ \dfrac{1}{r} & 1 & \dfrac{z}{r} & 0 & 0 & 0 \\ 0 & 0 & 1 & 0 & 1 & 0 \end{bmatrix} \begin{Bmatrix} a_1 \\ a_2 \\ a_3 \\ a_4 \\ a_5 \\ a_6 \end{Bmatrix} \qquad \text{...(17)}$$

Substituting equations (10) and (11) into equation (17) and making use of equation (a), we obtain:

$$\{\varepsilon\} = \frac{1}{2A} \begin{bmatrix} \beta_i & 0 & \beta_j & 0 \\ 0 & \gamma_i & 0 & \gamma_j \\ \dfrac{\alpha_i}{r} + \beta_i + \dfrac{\gamma_i z}{r} & 0 & \dfrac{\alpha_j}{r} + \beta_j + \dfrac{\gamma_j z}{r} & 0 & \dfrac{\alpha_m}{r} \\ \gamma_i & \beta_i & \gamma_j & \beta_j \end{bmatrix}$$

$$\begin{bmatrix} \beta_m & 0 \\ 0 & \gamma_m \\ +\beta_m + \dfrac{\gamma_m z}{r} & 0 \\ \gamma_m & \beta_m \end{bmatrix} \begin{Bmatrix} u_i \\ w_i \\ u_j \\ w_j \\ u_m \\ w_m \end{Bmatrix} \qquad \text{...(18)}$$

Or, rewriting equation(18) in simplified matrix form:

$$\{\varepsilon\}=\begin{bmatrix}\underline{B}_i & \underline{B}_j & \underline{B}_m\end{bmatrix}\begin{Bmatrix} u_i \\ w_i \\ u_j \\ w_j \\ u_m \\ w_m \end{Bmatrix} \qquad \ldots(19)$$

Where, $[\underline{B}_i]=\dfrac{1}{2A}\begin{bmatrix} \beta_i & 0 \\ 0 & \gamma_i \\ \dfrac{\alpha_i}{r}+\beta_i+\dfrac{\gamma_i z}{r} & 0 \\ \gamma_i & \beta_i \end{bmatrix} \qquad \ldots(20)$

Similarly, we obtain sub-matrices $\underline{B}_j$ and $\underline{B}_m$ by replacing the subscript i with j and then with m in equation (20). Rewriting equation (19) in compact matrix form, we have:

$$\{\varepsilon\}=[B]\;\{d\} \qquad \ldots(21)$$

Where, $[B]=\begin{bmatrix}\underline{B}_i & \underline{B}_j & \underline{B}_m\end{bmatrix}$

Note that [B] is a function of the r and z coordinates. Therefore, in general, the strain ε_θ will not be constant.

The stresses are given by:

$$\{\sigma\}=[D]\;[B]\;\{d\} \qquad \ldots(23)$$

Where [D] is given by the first matrix on the right side of equation (6)

Step 4: Derive the Element Stiffness Matrix and Equations

The stiffness matrix is:

$$[k]=\iiint_V [B]^T\;[D][B]dV \qquad \ldots(24)$$

Or $\quad [k]=2\pi\iint_A [B]^T\;[D][B]r\,dr\,dz \qquad \ldots(25)$

After integrating along the circumferential boundary. The [B] matrix, equation (22), is a function of r and z. Therefore, [k] is a function of r and z and is of order 6 x 6.

We can evaluate equation (25) for [k] by one of three methods:

- Numerical integration (Gaussian quadrature).

- Explicit multiplication and term-by-term integration.

- Evaluate [B] for a centroid point $(\bar{r},\bar{z})$ of the element.

$$\bar{r}=\bar{r}=\frac{r_i+r_j+r_m}{3} \qquad \bar{z}=\bar{z}=\frac{z_i+z_j+z_m}{3} \qquad ...(26)$$

And define $\left[B\left((\bar{r},\bar{z})\right)\right]=\left[\bar{B}\right]$. Therefore, as a first approximation:

$$[k]=2\pi\bar{r}\,A\left[\bar{B}\right]^{T}[D]\left[\bar{B}\right] \qquad ...(27)$$

If the triangular subdivisions are consistent with the final stress distribution (that is, small elements in regions of high stress gradients).

Distributed Body Forces

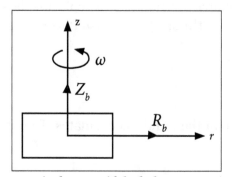

(c) Axisymmetric element with body forces per unit volume.

Loads such as gravity (in the direction of the z axis) or centrifugal forces in rotating machine parts (in the direction of the r axis) are considered to be body forces (as shown in the figure(c)). The body forces can be found by:

$$\{f_b\}=2\pi\iint_{A}[N]^{T}\begin{Bmatrix}R_b\\Z_b\end{Bmatrix}r\,dr\,dz \qquad ...(28)$$

Where, $R_b = \omega^2\rho r$ for a machine part moving with a constant angular velocity ω about the z axis, with material mass density ρ and radial coordinates and where, Z_b is the body force per unit volume due to the force of gravity.

Considering the body force at node i, we have:

$$\{f_{bi}\}=2\pi\iint_{A}[N_i]^{T}\begin{Bmatrix}R_b\\Z_b\end{Bmatrix}r\,dr\,dz \qquad ...(29)$$

Where,

Multiplying and integrating in equation (29), we obtain:

$$\{f_{bi}\} = \frac{2\pi}{3}\begin{Bmatrix} \overline{R}_b \\ \overline{Z}_b \end{Bmatrix} A\overline{r} \qquad \qquad ...(31)$$

Where, the origin of the coordinates has been taken as the centroid of the element and is the radially directed body force per unit volume evaluated at the centroid of the element. The body forces at nodes j and m are identical to those given by equation (31) for node i. Hence, for an element, we have:

$$\{f_b\} = \frac{2\pi\overline{r}A}{3}\begin{Bmatrix} \overline{R}_b \\ Z_b \\ \overline{R}_b \\ Z_b \\ \overline{R}_b \\ Z_b \end{Bmatrix} \qquad \qquad ...(32)$$

Where, $\overline{R}_b = \omega^2 \rho \overline{r}$ ...(33)

Equation (32) is a first approximation to the radially directed body force distribution.

Surface Forces

Surface forces can be found by:

$$\{f_s\} = \iint_S [N_s]^T \{T\} dS \qquad \qquad ...(34)$$

Where again [Ns] denotes the shape function matrix evaluated along the surface where the surface traction acts. For radial and axial pressures p_r and p_z, respectively, we have:

$$\{f_s\} = \iint_S [N_s]^T \begin{Bmatrix} p_r \\ p_z \end{Bmatrix} dS \qquad \qquad ...(35)$$

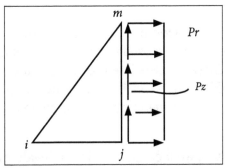

(d) Axisymmetric element with surface forces.

For example, along the vertical face j_m of an element, let uniform loads p_r and p_z be applied, as shown in the figure (d) along surface $r = r_j$. We can use equation (35) written for each node separately. For instance, for node j, substituting Nj from equations (13) into equation (35), we have,

$$\{f_{sj}\} = \int_{z_j}^{z_m} \frac{1}{2A} \begin{bmatrix} \alpha_j + \beta_j r + \gamma_j z & 0 \\ 0 & \alpha_j + \beta_j r + \gamma_j z \end{bmatrix} \begin{Bmatrix} p_r \\ p_z \end{Bmatrix} 2\pi r_j dz$$

evaluated at $r = r_i, z = z$ ...(36)

Performing the integration of (36) explicitly, along with similar evaluations for fsi and fsm, we obtain the total distribution of surface force to nodes i, j and m as:

$$\{f_s\} = \frac{2\pi r_j (z_m - z_j)}{2} \begin{Bmatrix} 0 \\ 0 \\ p_r \\ p_z \\ p_r \\ p_z \end{Bmatrix} \qquad ...(37)$$

Steps 5–7: Derivation of the Stiffness Matrix

Steps 5–7, which involve assembling the total stiffness matrix, total force matrix and total set of equations, solving for the nodal degrees of freedom and calculating the element stresses, are analogous for the CST element, except the stresses are not constant in each element.

They are usually determined by one of two methods that we use to determine the LST element stresses. Either we determine the centroid element stresses or we determine the nodal stresses for the element and then average them.

3.6   Numerical Integration: One Dimensional

The integrations, we generally encounter in finite element methods, are quite complicated and it is not possible to find a closed form solutions to those problems. Exact and explicit evaluation of the integral associated to the element matrices and the loading vector is not always possible because of the algebraic complexity of the coefficient of the different equation (i.e., the stiffness influence coefficients, elasticity matrix, loading functions etc.).

In the finite element analysis, we face the problem of evaluating the following types of integrations in one, two and three dimensional cases respectively. These are necessary to compute element stiffness and element load vector.

$$\int \phi(\xi)d\xi; \qquad \int \phi(\xi,\eta)d\xi d\eta; \qquad \int \phi(\xi,\eta,\zeta)d\xi d\eta d\zeta;$$

Approximate solutions to such problems are possible using certain numerical techniques. Several numerical techniques are available, in mathematics for solving definite integration problems, including, mid-point rule, trapezoidal-rule, Simpson's 1/3rd rule, Simpson's 3/8th rule and Gauss Quadrature formula. Among these, Gauss Quadrature technique is most useful one for solving problems in finite element method.

Gauss Quadrature for One-Dimensional Integrals

The concept of Gauss Quadrature is first illustrated in one dimension in the context of an integral in the form of $I = \int_{-1}^{+1}\phi(\xi)d\xi$ from $\int_{x_1}^{x_2}f(x)dx$. To transform from an arbitrary interval of $x_1 \le x \le x_2$ to an interval of $-1 \le \xi \le 1$, we need to change the integration function from $f(x)$ to $\phi(\xi)$ accordingly.

Thus, for a linear variation in one dimension, one can write the following relations:

$$x = \frac{1-\xi}{2}x_1 + \frac{1+\xi}{2}x_2 = N_1 x_1 + N_2 x_2$$

$$\text{so for } \xi = -1, \ x = \frac{-1(-1)}{2}x_1 + \frac{1-1}{2}x_2 = x_1$$

$$\xi = +1, \ x = x_2$$

$$\therefore I = \int_{x_1}^{x_2}f(x)dx = \int_{-1}^{+1}\phi(\xi)d\xi$$

Numerical integration based on Gauss Quadrature assumes that the function $\phi(\xi)$ will be evaluated over an interval $-1 \le \xi \le 1$. Considering an one-dimensional integral, Gauss Quadrature represents the integral $\phi(\xi)$ in the form of:

$$I = \int_{-1}^{+1}\phi(\xi)d\xi \approx \sum_{i=1}^{n}w_i\phi(\xi_i) \approx w_1\phi(\xi_1) + w_2\phi(\xi_2) + \ldots + w_n\phi(\xi_n) \qquad \ldots(1)$$

Where, the $\xi_1, \xi_2, \xi_3, \ldots, \xi_n$ represents n numbers of points known as Gauss Points and the corresponding coefficients $w_1, w_2, w_3, \ldots, w_n$ are known as weights. The location and weight coefficients of Gauss points are calculated by Legendre polynomials. Hence this method is also sometimes referred as Gauss-Legendre Quadrature method.

The summation of these values at n sampling points gives the exact solution of a polynomial integrand of an order up to 2n-1. For example, considering sampling at two

Gauss points we can get exact solution for a polynomial of an order (2×2-1) or 3. The use of more number of Gauss points has no effect on accuracy of results but takes more computation time.

One-Point Formula

Considering n = 1, equation (1) can be written as:

$$\int_{-1}^{1} \phi(\xi)d\xi \approx w_1\, \phi(\xi_1) \qquad\qquad ...(2)$$

Since there are two parameters w_1 and ξ_1, we need a first order polynomial for $\phi(\xi)$ to evaluate the equation (2) exactly. For example, considering, $\phi(\xi) = a_0 + a_1\xi$.

$$\text{Error } = \int_{-1}^{1}(a_0 + a_1\,\xi)d\xi - w_1\,\phi(\xi_1) = 0$$

$$\Rightarrow 2a_0 - w_1(a_0 + a_1\xi_1) = 0$$

$$\Rightarrow a_0(2 - w_1) - w_1 a_1 \xi_1 = 0$$

Thus, the error will be zero if w1= 2 and ξ1 = x=. Putting these in equation (2), for any general φ, we have:

$$I = \int_{-1}^{1}\phi(\xi)d\xi = 2\phi(0)$$

This is exactly similar to the well-known midpoint rule.

Two-Point Formula

If we consider n = 2, then the equation (1) can be written as:

$$\int_{-1}^{1}\phi(\xi)d\xi \approx w_1\,\phi(\xi_1) + w_2\phi(\xi_2) \qquad\qquad ...(3)$$

This means we have four parameters to evaluate. Hence we need a 3rd order polynomial for φ(ξ) to exactly evaluate equation.(3). Considering $\phi(\xi) = a_0 + a_1\,\xi + a_2\,\xi^2 + a_3\,\xi^3$:

$$\text{Error } = \left[\int_{-1}^{1}(a_0 + a_1\,\xi + a_2\,\xi^2 + a_3\,\xi^3)d\xi\right] - \left[w_1\,\phi(\xi_1) + w_2\phi(\xi_2)\right]$$

$$\Rightarrow 2a_0 + \frac{2}{3}a_2 - w_1\left(a_0 + a_1\,\xi_1 + a_2\,\xi_1^2 + a_3\,\xi_1^3\right) - w_2\left(a_0 + a_1\,\xi_2 + a_2\,\xi_2^2 + a_3\,\xi_2^3\right) = 0$$

$$\Rightarrow (2 - w_1 - w_2)a_0 - (w_1\,\xi_1 + w_2\,\xi_2)$$

$$a_1 + \left(\frac{2}{3} - w_1\,\xi_1^2 - w_2\,\xi_2^2\right)a_2 - \left(w_1\,\xi_1^3 - w_2\,\xi_2^3\right)a_3 = 0$$

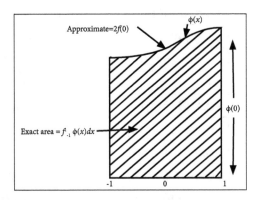

Requiring zero error yield:

$$w_1 + w_2 = 2$$

$$w_1\,\xi_1 + w_2\,\xi_2 = 0$$

$$w_1\,\xi_1^2 + w_2\,\xi_2^2 = \frac{2}{3}$$

$$w_1\,\xi_1^3 + w_2\,\xi_2^3 = 0$$

These nonlinear equations have the unique solution as:

$$w_1 = w_2 = 1 \qquad\qquad \xi_1 = -\xi_2 = -1/\sqrt{3} = -0.5773502691$$

From this solution, we can conclude that n-point Gaussian Quadrature will provide an exact solution if $\phi\,(\xi)$ is a polynomial of order (2n-1) or less. Table gives the values of w1 and ξx for Gauss Quadrature formulas of orders n = 1 through n = 6. From the table it can be observed that the gauss 59 points are symmetrically placed with respect to origin and those symmetrical points have the same weights.

For accuracy in the calculation maximum number digits for gauss point and gauss weights should be taken. The Location and weights given in the table must be used when the limits of integration ranges from -1 to 1. Integration limits other than [-1, 1], should be appropriately changed to [-1, 1] before applying these values.

Gauss Points and Corresponding Weights

Number of gauss point, n	Gauss point location, ξi	Weight w_i
1	0.0	2.0

2	±0.5773502692 (= ± 1√3)	1.0
3	0.0	0.8888888889 (=8/9)
	±0.7745966692 (= ± 1√0.6)	0.5555555556 (=5/9)
4	±0.3399810436	0.6521451549
	±0.861363116	0.3478548451
5	0.0	0.5688888889
	±0.5384693101	0.4786286705
	±0.9061798459	0.2369268851
6	±0.2386191861	0.4679139346
	±0.6612093865	0.3607615730
	±0.9324695142	0.1713244924

3.6.1 Two and Three Dimensional

Gauss Quadrature for Two-Dimensional Integrals

For two dimensional integration problems the above mentioned method can be extended by first evaluating the inner integral, keeping η constant and then evaluating the outer integral. Thus:

$$I = \int_{-1}^{1}\int_{-1}^{1} \phi(\xi,\eta)\,d\xi\,d\eta \approx \int_{-1}^{1}\left[\sum_{i=1}^{n} w_i\,\phi(\xi_i,\eta)\right]d\eta \approx \sum_{i=1}^{n} w_j\left[\sum_{i=1}^{n} w_i\,\phi(\xi_i,\eta_j)\right]$$

Or,

$$I \approx \sum_{i=1}^{n}\sum_{j=1}^{n} w_i\,w_j\,\phi(\xi_i,\eta_j)$$

In a matrix form we can rewrite the above expression as:

$$I \approx \begin{bmatrix} w_1 & w_2 & \cdots & w_n \end{bmatrix} \begin{bmatrix} \phi(\xi_1,\eta_1) & \phi(\xi_1,\eta_2) & & \phi(\xi_1,\eta_n) \\ \phi(\xi_2,\eta_1) & \phi(\xi_2,\eta_2) & & \phi(\xi_2,\eta_n) \\ & & \ddots & \\ \phi(\xi_n,\eta_1) & \phi(\xi_n,\eta_2) & & \phi(\xi_n,\eta_n) \end{bmatrix} \begin{Bmatrix} w_1 \\ w_2 \\ \vdots \\ w_n \end{Bmatrix}$$

Example: Evaluate the integral: $I = \int_{y=c=-4}^{y=d=4}\int_{x=a=2}^{x=b=3} (1-x)^2(2-y)^2\,dx\,dy$

Solution:

Before applying the Gauss Quadrature formula, the above integral should be converted in terms of ξ and η the existing limits of y should be changed from [-4,4] to [-1, 1] and that of ξ is from [2,3] to [-1,1].

$$x = \frac{(b-a)}{2}\xi + \frac{(b+a)}{2} = \frac{(\xi+5)}{2}; \quad dx = \frac{d\xi}{2}$$

$$y = \frac{(d-c)}{2}\eta + \frac{(d+c)}{2} = 4\eta; \qquad dy = 4d\eta$$

$$I = 2\int_{\eta=-1}^{\eta=+1}\int_{\xi=-1}^{\xi=+1}\left(\frac{3+\xi}{2}\right)^2 (2-4\eta)^2\,d\xi\,d\eta = \int_{\eta=-1}^{\eta=+1}\int_{\xi=-1}^{\xi=+1} \phi(\xi,\eta)\,d\xi\,d\eta$$

Where,

$$\phi(\xi,\eta) = 2\left(\frac{3+\xi}{2}\right)^2 (2-4\eta)^2 = 2(3+\xi)^2(1-2\eta)^2$$

Gauss points for two-dimensional integral.

$$\xi_1 = -\frac{1}{\sqrt{3}}; \eta_1 = -\frac{1}{\sqrt{3}}; \xi_2 = \frac{1}{\sqrt{3}}; \eta_2 = \frac{1}{\sqrt{3}}$$

$$\phi(\xi_1,\eta_1) = 2\left(3-\frac{1}{\sqrt{3}}\right)^2\left(1+\frac{2}{\sqrt{3}}\right)^2 = 54.49857$$

$$\phi(\xi_2,\eta_1) = \left(\frac{3+\frac{1}{\sqrt{3}}}{2}\right)^2\left(2+\frac{4}{\sqrt{3}}\right)^2 = 118.83018$$

$$\phi(\xi_2,\eta_2) = \left(\frac{3+\frac{1}{\sqrt{3}}}{2}\right)^2\left(2-\frac{4}{\sqrt{3}}\right)^2 = 0.61254$$

$$\phi(\xi_1, \eta_2) = \left(\frac{3 - \frac{1}{\sqrt{3}}}{2}\right)^2 \left(2 - \frac{4}{\sqrt{3}}\right)^2 = 0.28093$$

$$I = \{w_1 \quad w_2\} \begin{bmatrix} \phi(\xi_1, \eta_1) & \phi(\xi_2, \eta_2) \\ \phi(\xi_2, \eta_1) & \phi(\xi_2, \eta_2) \end{bmatrix} \begin{Bmatrix} w_1 \\ w_2 \end{Bmatrix}$$

$$= \{1 \quad 1\} \begin{bmatrix} 54.49857 & 0.28093 \\ 118.83018 & 0.61254 \end{bmatrix} \begin{Bmatrix} 1 \\ 1 \end{Bmatrix}$$

$$= 174.22222 \text{ agrees with the exact value } 174.22222$$

Gauss Quadrature for Three-Dimensional Integrals

In a similar way one can extend the gauss Quadrature for three dimensional problems also and the integral can be expressed by:

$$I = \int_{-1}^{1} \int_{-1}^{1} \int_{-1}^{1} \phi(\xi, \eta, \zeta) d\xi \, d\eta \, d\zeta \approx \sum_{i=1}^{n} \sum_{j=1}^{n} \sum_{k=1}^{n} w_i \, w_j \, w_k \, \phi(\xi_i, \eta_j, \zeta_k)$$

The above equation will produce exact value for a polynomial integrand if the sampling points are selected as described the problems.

Problem

1. Numerical integration for two dimensional problems. Evaluate the integral:

$$I = \int_{-2}^{3} (x^2 + 11x - 32) dx$$ Using one, two and three point gauss Quadrature. Also, find the exact solution for comparison of accuracy.

Solution:

The existing limits of integration should be changed from [-2, +3] to [-1, +1]. Assuming, $\xi = a + bx$, the upper and lower limit can be changed. i.e., at x1 = -2, $\xi_1 = -1$ and at x2 = 3, $\xi_2 = +1$. Thus, putting these limits and solving for a & b, we get a = -0.2 and b = 0.4. The relation between two coordinate systems will become:

$$\xi = -0.2 + 0.4x \text{ or } x = \frac{5\xi + 1}{2} \text{ and } dx = 2.5 d\xi$$

Thus, the initial equation can be written as:

$$I = \int_{-2}^{3} (x^2 + 11x - 32) dx = 2.5 \int_{-1}^{+1} \left[\left(\frac{5\xi + 1}{2}\right)^2 + 11\left(\frac{5\xi + 1}{2}\right) - 32 \right] d\xi$$

(i) Exact solution:

$$I = \int_{-2}^{3} \left(x^2 + 11x - 32 \right) dx$$

$$= \left[\frac{x^3}{3} + \frac{11x^2}{2} - 32x \right]_{-2}^{3}$$

$$= \left[9 + \frac{99}{2} - 96 \right] - \left[-\frac{8}{3} + 22 + 64 \right]$$

$$= -37.5 - 83.333333 = -120.83333$$

Thus, $I_{exact} = -120.83333$

(ii) One point formula:

$$I = \int_{-1}^{+1} \phi(\xi) d\xi = w_1 \phi(\xi_1)$$

For one point formula in Gauss Quadrature integration, w1=2, ξ1=0. Thus:

$$I_1 = 2 \times 2.5 \left[\left(\frac{5 \times 0 + 1}{2} \right)^2 + 11 \left(\frac{5 \times 0 + 1}{2} \right) - 32 \right]$$

$$= 5 \left[\frac{1}{4} + \frac{11}{2} - 32 \right] = -131.25$$

Thus, % of error $= (120.83333 - 131.25) \times 100 / 120.83333 = 8.62\%$.

(iii) Two Point Formula:

Here, for two point formula in Gauss Quadrature integration:

$$w_1 = w_2 = 1.0 \text{ and } \xi_1 = -\xi_2 = -\frac{1}{\sqrt{3}}. \text{ Thus,}$$

$$I_2 = w_1 \phi(\xi_1) + w_2 \phi(\xi_2)$$

$$1.0 \times 2.5 \times \left[\left(\frac{\frac{-5}{\sqrt{3}}+1}{2} \right)^2 + 11\left(\frac{\frac{-5}{\sqrt{3}}+1}{2} \right) - 32 \right] + 1.0 \times 2.5 \times \left[\left(\frac{\frac{5}{\sqrt{3}}+1}{2} \right)^2 + 11\left(\frac{\frac{5}{\sqrt{3}}+1}{2} \right) - 32 \right]$$

$$= (0.88996 - 10.37713 - 32) \times 2.5 + (3.77671 + 21.3771 - 32) \times 2.5$$

$$=-48.3333\times2.5$$

$$=-120.83325$$

Thus, % of error $= (120.83333-120.83325)\times100/120.83333 = 6.62\times10^{-05}$

(iv) Three Point Formula:

Here, for three point formula in Gauss Quadrature integration:

$w_1=0.8889,\qquad \xi_{51}=0.0$

$w_2=0.5556,\qquad \xi_{52}=+0.7746$

$w_3=0.5556,\qquad \xi_{53}=-0.7746$

Thus,

$$I_3=w_1\,\phi(\xi_1)+w_2\,\phi(\xi_2)+w_3\,\phi(\xi_3)$$

$$I_3=0.8889\times2.5\times\left[\left(\frac{5\times0+1}{2}\right)^2+11\times\frac{5\times0+1}{2}-32\right]$$

$$+0.5556\times2.5\times\left[\left(\frac{5\times0.7746+1}{2}\right)^2+11\times\frac{5\times0.7746+1}{2}-32\right]$$

$$+0.55556\times2.5\times\left[\left(\frac{-5\times0.7746+1}{2}\right)^2+11\times\frac{-5\times0.7746+1}{2}-32\right]$$

$$I_3=0.8889\times2.5\times[0.25+5.5-32]$$

$$+0.5556\times2.5\times[5.9365+26.8015-32]$$

$$+0.5556\times2.5\times[2.0635-15.8015-32]$$

$$=2.5\times(-23.3336+0.4100-25.4120)$$

$$=-2.5\times48.3356=-120.839$$

Thus, % of error = (120.83333-120.839)×100/120.83333 = 4.69×10-03. However, difference of results will approach to zero, if few more digits after decimal points are taken in calculation.

2. Numerical integration for three dimensional problems: Evaluate the integral:

$$I = \int_{-1}^{1}\int_{-1}^{1}\int_{-1}^{1}(1-2\xi)^2(1-\eta)^2(3\zeta-2)^2\,d\xi\,d\eta\,d\zeta$$

Solution:

Using two point gauss Quadrature formula for the evaluation of three dimensional integration, we have the following sampling points and weights:

$$w_1 = w_2 = 1$$

$$\xi_1 = -0.5773502692$$

$$\xi_2 = 0.5773502692$$

$$\eta_1 = -0.5773502692$$

$$\eta_2 = 0.5773502692$$

$$\zeta_1 = -0.5773502692$$

$$\zeta_2 = 0.5773502692$$

Putting the above the values in $\phi(\xi,\eta,\zeta) = (1-2\xi)^2(1-\eta)^2(3\zeta-2)^2$ one can find the values in 8 (2x2x2) sampling point:

$$\phi(\xi_1,\eta_1,\zeta_1) = 160.8886$$

$$\phi(\xi_1,\eta_1,\zeta_2) = 0.8293$$

$$\phi(\xi_1,\eta_2,\zeta_1) = 11.5513$$

$$\phi(\xi_1,\eta_2,\zeta_2) = 0.0595$$

$$\phi(\xi_2,\eta_1,\zeta_1) = 0.8293$$

$$\phi(\xi_2,\eta_1,\zeta_2) = 0.0043$$

$\phi(\xi_2, \eta_2, \zeta_1) = 0.0595$

$\phi(\xi_2, \eta_2, \zeta_2) = 0.0003$

Now, $I = \sum\limits_{i=1}^{2}\sum\limits_{j=1}^{2}\sum\limits_{k=1}^{2} w_i \, w_j \, w_k \; \phi(\xi_i, \eta_j, \zeta_k)$

Thus, $I = w_1 \, w_1 \, w_1 \, \phi(\xi_1, \eta_1, \zeta_1) + w_1 \, w_1 \, w_2 \, \phi(\xi_1, \eta_1, \zeta_2) + \ldots$

$+ w_2 \, w_2 \, w_2 \, \phi(\xi_2, \eta_2, \zeta_2) = 174.222$, where,

as $I_{exact} = 174.222$.

Analysis of Frame Structures and Additional Applications of FEM

4.1 Stiffness of Truss Members

A truss is one of the simplest and most widely used structural members. It is a straight bar that is designed to take only axial forces, therefore it deforms only in its axial direction.

In planar trusses there are two components in the x and y directions for the displacements as well as for the forces.

For space trusses, however, there will be three components in the x, y and z directions for both displacements and forces. In skeletal structures consisting of truss members, the truss members are joined together by pins or hinges (not by welding), so that there are only forces (not moments) transmitted between the bars.

Types of Truss Element

- Plane truss
- Space truss

Plane Truss

Members lie in a single plane.

FEM allows to remove the rigid body restriction and to solve statically indeterminate truss problems.

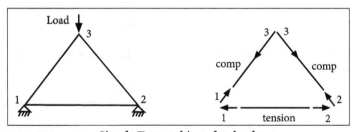

Simple Truss subjected to load.

Finite Element Formulation

$$\text{Average stress } \sigma = \frac{F}{A}$$

$$\text{Average strain } \varepsilon = \frac{\Delta L}{L}$$

In elastic region,

$$\sigma = E\,\varepsilon \; (\text{ Hook's law })$$

$$F = \left(\frac{AE}{L}\right)\Delta L$$

Element stiffness matrix

$$K_{eq} = \frac{AE}{L}$$

Global displacement and local displacement are related:

$$u_{ix.} = u_{ix.}\cos\theta - u_{iy.}\sin\theta$$

$$u_{iy.} = u_{ix.}\sin\theta + u_{iy.}\cos\theta$$

$$u_{jx.} = u_{jx.}\cos\theta - u_{jy.}\sin\theta$$

$$u_{jy.} = u_{jx.}\sin\theta + u_{jy.}\cos\theta$$

Similarly global and local forces are related:

$$u = [T]u$$

$$F = [T]f$$

Space Truss

Three dimensional truss often known as space truss:

- Simple space truss has six members joined together at their ends to form a tetrahedron.

- Complex structure can be created by adding new members to simple truss.

- Finite element formulation of space trusses is an extension of analysis of plane truss.

Analysis

Analysis of frame structures can be carried out by the approach of stiffness method. However, such types of structures can also be analyzed by finite element method.

A unified formulation will be demonstrated based on finite element concept in this module for the analysis of frame like structures. A truss structure is composed of slender members pin jointed together at their end points.

Truss element can resist only axial forces (tension or compression) and can deform only in its axial direction. Therefore, in case of a planar truss, each node has components of displacements parallel to X and Y axis.

Planar trusses lie in a single plane and are used to support roofs and bridges. Such members will not be able to carry transverse load or bending moment. The major benefits of use of truss structures are lightweight, re-construct able, reconfigurable and mobile. Configuration of few standard truss structures are shown in the figure (1).

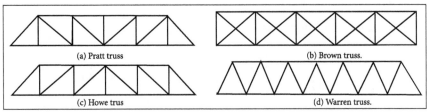

(1) Configuration of various truss structures.

Element Stiffness of a Truss Member

Since, the truss is an axial force resisting member, the displacement along its axis only will be developed due to axial load. Therefore, using Pascal's triangle, the displacement function of truss member for development of shape function can be expressed as:

$$u(x)=\alpha_0+\alpha_1 x=\begin{bmatrix}1 & x\end{bmatrix}\begin{Bmatrix}\alpha_0\\\alpha_1\end{Bmatrix} \qquad ...(1)$$

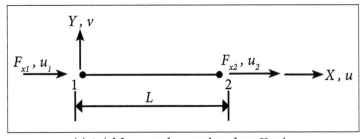

(a) Axial force on the member along X axis.

Applying boundary conditions as shown in the figure (a),

At x = 0, $u(0)=u_1$ and at x = L, $u(L)=u_2$

Thus, $a_0 = u_1$ and $\alpha_1 = \dfrac{u_2 - u_1}{L}$. Therefore,

$$u(x) = \left(1 - \frac{x}{L}\right)u_1 + \frac{x}{L}u_2 = [N]\{u\} \qquad ...(2)$$

Here, N is the shape function of the element and is expressed as:

$$[N] = \left[1 - \frac{x}{L} \quad \frac{x}{L}\right] \qquad ...(3)$$

So we get the element stiffness matrix as:

$$[k] = \iiint_\Omega [B]^T [D][B]\, d\Omega \qquad ...(4)$$

Where, $[B] = \dfrac{d[N]}{dx} = \left[-\dfrac{1}{L} \quad \dfrac{1}{L}\right]$

So, the stiffness matrix will become:

$$= \int_0^L [B]^T E[B] A\, dx = A dx = AE \int_0^L \left\{\begin{array}{c} -\dfrac{1}{L} \\ \dfrac{1}{L} \end{array}\right\} \left[-\dfrac{1}{L} \quad \dfrac{1}{L}\right] dx = \frac{AE}{L}\begin{bmatrix} 1 & -1 \\ -1 & 1 \end{bmatrix}$$

Thus, the stiffness matrix of the truss member along its member axis will be:

$$[k] = \frac{AE}{L}\begin{bmatrix} 1 & -1 \\ -1 & 1 \end{bmatrix} \qquad ...(5)$$

Element Stiffness of Truss Member with Varying Cross Section

Now, let us find the stiffness matrix of a pin-jointed member of length L with respect to local axis, having cross sectional areas A1 and A2 at the two ends of the member as shown in the figure below:

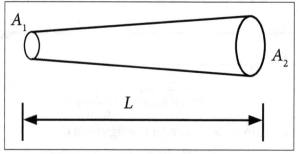

(b) Member with varying cross section.

From the above figure, the cross sectional area at a distance of x from left end can be expressed as:

$$A_x = A_1 + \frac{A_2 - A_1}{L}x \qquad \ldots(6)$$

As it is a pin-jointed member, the displacement at any point may be expressed in terms of nodal displacement as $u = N_1 u_1 + N_2 u_2$. Similarly the cross sectional area at any point may be represented in terms of the cross sectional area of the two ends. Thus $A_x = N_1 A_1 + N_2 A_2$

Where, the shape functions are: $N_1 = 1 - \frac{x}{L}$; $N_2 = \frac{x}{L}$

Now, the strain may be written as:

$$\varepsilon_x = \frac{\partial u}{\partial x} = \frac{\partial N_1}{\partial x} u_1 + \frac{\partial N_2}{\partial x} u_2 = -\frac{1}{L} u_1$$

$$+ \frac{1}{L} u_2 = \frac{1}{L}\begin{bmatrix} -1 & 1 \end{bmatrix}\begin{Bmatrix} u_1 \\ u_2 \end{Bmatrix} = [B]\{u\} \qquad \ldots(7)$$

As the stress is proportional to strain according to Hook's law, the stress-strain relationship will be as follows:

$$\sigma_x = E\varepsilon_x = \frac{E}{L}\begin{bmatrix} -1 & 1 \end{bmatrix}\begin{Bmatrix} u_1 \\ u_2 \end{Bmatrix} = E[B][\{u\}] \qquad \ldots(8)$$

Now the strain energy may be expressed as:

$$U = \frac{1}{2}\int_V \varepsilon_x^T \sigma_x \, dv = \frac{1}{2}\int_0^L \varepsilon_x^T E\varepsilon_x A_x \, dx = \frac{1}{2}\int_0^L \{u\}^T [B]^T E[B]\{u\} A_x dx \qquad \ldots(9)$$

Applying Castiglione's theorem, the force will become:

$$\{F\} = \frac{\partial U}{\partial \{u\}} = \int_0^L [B]^T E[B]\{u\} A_x d_x$$

$$= \frac{E}{L^2}\int_0^L \begin{bmatrix} -1 & 1 \end{bmatrix}^T \begin{bmatrix} -1 & 1 \end{bmatrix} A_x dx \begin{Bmatrix} u_1 \\ u_2 \end{Bmatrix} = [k]\{d\} \qquad \ldots(10)$$

Thus, the stiffness matrix will be:

$$[k] = \frac{E}{L^2}\begin{bmatrix} 1 & -1 \\ -1 & 1 \end{bmatrix}\int_0^L \left(A_1 x + \frac{A_2 - A_1}{2L}x\right)dx = \frac{E}{L^2}\begin{bmatrix} 1 & -1 \\ -1 & 1 \end{bmatrix}\left[A_1 x + \frac{A_2 - A_1}{2L}x^2\right]_0^L$$

$$= \frac{E}{L} \begin{bmatrix} 1 & -1 \\ -1 & 1 \end{bmatrix} \left[A_1 + \frac{A_2 - A_1}{2} \right] = \frac{E}{2L}(A_1 + A_2) \begin{bmatrix} 1 & -1 \\ -1 & 1 \end{bmatrix} \qquad ...(11)$$

Generalized Stiffness Matrix of a Plane Truss Member

Let us consider a member making an angle 'θ' with X axis as shown in the figure below. By resolving the forces along local X and Y direction, the following relations are obtained:

$$\left. \begin{aligned} \overline{F}_{x1} &= F_{x1} \cos\theta + F_{y1} \sin\theta \\ \overline{F}_{x2} &= F_{x2} \cos\theta + F_{y2} \sin\theta \\ \overline{F}_{y1} &= -F_{x1} \sin\theta + F_{y1} \cos\theta \\ \overline{F}_{y2} &= -F_{x2} \sin\theta + F_{y2} \cos\theta \end{aligned} \right\} \qquad ...(12)$$

Where, $\overline{F}_{x1}$ and $\overline{F}_{x2}$ are the axial forces along the member axis $\overline{X}$. Similarly, $\overline{F}_{y1}$ and $\overline{F}_{y2}$ are the forces perpendicular to the member axis $\overline{X}$.

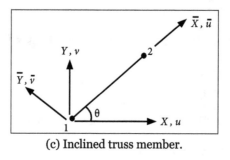

(c) Inclined truss member.

The relationship expressed in equation (12) can be rewritten in matrix form as follows:

$$\begin{Bmatrix} \overline{F}_{x1} \\ \overline{F}_{y1} \\ \overline{F}_{x2} \\ \overline{F}_{y2} \end{Bmatrix} = \begin{bmatrix} \cos\theta & \sin\theta & 0 & 0 \\ -\sin\theta & \cos\theta & 0 & 0 \\ 0 & 0 & \cos\theta & \sin\theta \\ 0 & 0 & -\sin\theta & \cos\theta \end{bmatrix} \begin{Bmatrix} F_{x1} \\ F_{y1} \\ F_{x2} \\ F_{y2} \end{Bmatrix} \qquad ...(13)$$

Now, the above equation can be expressed in short as:

$$\{\overline{F}\} = [T]\{F\} \qquad ...(14)$$

Here, [T] is called transformation matrix. This relates between the global (X, Y axis) and member axis ($\overline{X}, \overline{Y}$ axis). Similarly, the relations of nodal displacements between two coordinate systems may be written as:

$$\{\overline{d}\} = [T]\{d\} \qquad ...(15)$$

Again, the equation stated in (5) can be generalized and expressed with respect to the member axis including force and displacement vector as:

$$\begin{Bmatrix} \overline{F}_{x1} \\ \overline{F}_{y1} \\ \overline{F}_{x2} \\ \overline{F}_{y2} \end{Bmatrix} = \frac{AE}{L} \begin{bmatrix} 1 & 0 & -1 & 0 \\ 0 & 0 & 0 & 0 \\ -1 & 0 & 1 & 0 \\ 0 & 0 & 0 & 0 \end{bmatrix} \begin{Bmatrix} \overline{u}_1 \\ \overline{v}_1 \\ \overline{u}_2 \\ \overline{v}_2 \end{Bmatrix} \qquad \ldots(16)$$

Where, the nodal forces in direction are zero. The above equation may also be expressed in short as:

$$\{\overline{F}\} = [\overline{k}]\{\overline{d}\} \qquad \ldots(17)$$

Where, the matrices in the above equation are written with respect to the member axis. Now, equation (17) can be rewritten with the use of equation (14) and (15) as given below:

$$[T]\{F\} = [\overline{k}][T]\{d\} \qquad \ldots(18)$$

Or

$$\{F\} = [T]^{-1}[\overline{k}][T]\{d\} \qquad \ldots(19)$$

Here, the transformation matrix [T] is orthogonal, i.e., [T]-1 is equal to [T]T. Therefore, from the above relationship, the generalized stiffness matrix can be expressed as:

$$[k] = [T]^{T}[\overline{k}][T] \qquad \ldots(20)$$

Thus,

$$[k] = \begin{bmatrix} \cos\theta & -\sin\theta & 0 & 0 \\ \sin\theta & \cos\theta & 0 & 0 \\ 0 & 0 & \cos\theta & -\sin\theta \\ 0 & 0 & \sin\theta & \cos\theta \end{bmatrix}$$

$$\frac{AE}{L} \begin{bmatrix} 1 & 0 & -1 & 0 \\ 0 & 0 & 0 & 0 \\ -1 & 0 & 1 & 0 \\ 0 & 0 & 0 & 0 \end{bmatrix} \begin{bmatrix} \cos\theta & \sin\theta & 0 & 0 \\ -\sin\theta & \cos\theta & 0 & 0 \\ 0 & 0 & \cos\theta & \sin\theta \\ 0 & 0 & -\sin\theta & \cos\theta \end{bmatrix} \qquad \ldots.(21)$$

Or

$$[k] = \frac{AE}{L} \begin{bmatrix} \cos^2\theta & \sin\theta\cos\theta & -\cos^2\theta & -\sin\theta\cos\theta \\ \sin\theta\cos\theta & \sin^2\theta & -\sin\theta\cos\theta & -\sin^2\theta \\ -\cos^2\theta & -\sin\theta\cos\theta & \cos^2\theta & \sin\theta\cos\theta \\ -\sin\theta\cos\theta & -\sin^2\theta & \sin\theta\cos\theta & \sin^2\theta \end{bmatrix} \qquad \ldots(22)$$

The above stiffness matrix can be used for the analysis of two-dimensional truss problems.

4.2 Analysis of Truss

Element Stiffness of a 3-Node Truss Member

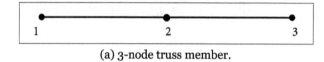

(a) 3-node truss member.

Here, the displacement function using Pascal's triangle can be expressed as:

$$u(x) = \alpha_0 + \alpha_1 x + \alpha_2 x^2 = \begin{bmatrix} 1 & x & x^2 \end{bmatrix} \begin{Bmatrix} \alpha_0 \\ \alpha_1 \\ \alpha_2 \end{Bmatrix} \qquad ...(1)$$

Applying boundary conditions:

At $x = 0$, $u(0) = u_1$, $x = L/2$, $u(L/2) = u_2$ and at $x = L$, $u(L) = u_3$

And solving for α_0, α_1 and α_2

$$\alpha_0 = u_1, \alpha_1 = \frac{-3u_1 + 4u_2 - u_3}{L} \text{ and } \alpha_2 = \frac{2u_1 - 4u_2 + 2u_3}{L^2}$$

Therefore,

$$u(x) = \left(1 - \frac{3x}{L} + \frac{2x^2}{L^2}\right) u_1 + \left[\frac{4x}{L} - \frac{4x^2}{L^2}\right] u_2$$

$$+ \left(-\frac{x}{L} + \frac{2x^2}{L^2}\right) u_3 = [N]\{u\} \qquad ...(2)$$

Here, N is the shape function of the element and is expressed as:

$$[N] = \left[\left(1 - \frac{3x}{L} + \frac{2x^2}{L^2}\right) \left(\frac{4x}{L} - \frac{4x^2}{L^2}\right) \left(-\frac{x}{L} + \frac{2x^2}{L^2}\right)\right] \qquad ...(3)$$

Now, the element stiffness matrix can be written as:

$$[k] = \iiint_\Omega [B]^T [D] [B] d\Omega \qquad ...(4)$$

Where, $[B]=\dfrac{d[N]}{dx}=\left[-\dfrac{3}{L}+\dfrac{4x}{L^2} \quad \dfrac{4}{L}-\dfrac{8x}{L^2} \quad -\dfrac{1}{L}+\dfrac{4x}{L^2}\right]$

$$[k]\iiint_{\Omega}[B]^{T}[D][B]d\Omega = \int_{0}^{L}[B]^{T}E[B]Adx$$

So, the stiffness matrix will be:

$$= AE \int_{0}^{L}\begin{Bmatrix} -\dfrac{3}{L}+\dfrac{4x}{L^2} \\[2mm] \dfrac{4}{L}-\dfrac{8x}{L^2} \\[2mm] -\dfrac{1}{L}+\dfrac{4x}{L^2} \end{Bmatrix} \times \left[-\dfrac{3}{L}+\dfrac{4x}{L^2} \quad \dfrac{4}{L}-\dfrac{8x}{L^2} \quad -\dfrac{1}{L}+\dfrac{4x}{L^2}\right]dx$$

$$=\dfrac{AE}{L^2}\int_{0}^{L}\begin{bmatrix} 9+\dfrac{16x^2}{L^2}-\dfrac{24x}{L} & -12+\dfrac{40x}{L}-\dfrac{32x^2}{L^2} & 3-\dfrac{16x}{L}+\dfrac{16x^2}{L^2} \\[3mm] -12+\dfrac{40x}{L}-\dfrac{32x^2}{L^2} & 16-\dfrac{64x}{L}+\dfrac{64x^2}{L^2} & -4+\dfrac{24x}{L}-\dfrac{32x^2}{L^2} \\[3mm] 3-\dfrac{16x}{L}+\dfrac{16x^2}{L^2} & -4+\dfrac{24x}{L}-\dfrac{32x^2}{L^2} & 1-\dfrac{8x}{L}+\dfrac{16x^2}{L^2} \end{bmatrix}dx \qquad ...(5)$$

After integrating the above equation, the stiffness matrix of the 3-node truss member will become:

$$[k]=\dfrac{AE}{3L}\begin{bmatrix} 7 & -8 & 1 \\ -8 & 16 & -8 \\ 1 & -8 & 7 \end{bmatrix} \qquad ...(6)$$

Problem:

1. Let us analyze the truss shown below by finite element method. Assume the cross sectional area of the inclined member as 1.5 times the area (A) of the horizontal and vertical members. Assume modulus of elasticity is constant for all the members and is E.

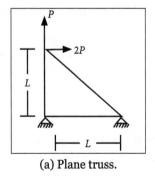

(a) Plane truss.

Solution:

Given:

Assume the cross sectional area of the inclined member as 1.5 times the area (A) of the horizontal and vertical members.

The analysis of truss starts with the numbering of members and joints as shown below:

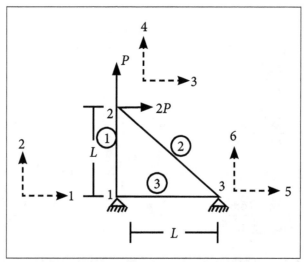

(b) Numbering of members and nodes.

The member information for the truss is shown in the table (a). The member and node numbers, modulus of elasticity, cross sectional areas are the necessary input data.

From the coordinate of the nodes of the respective members, the length of each member is computed. Here, the angle θ has been calculated considering anticlockwise direction. The signs of the direction cosines depend on the choice of numbering the nodal connectivity.

(a) Member Information for Truss

Member No.	Starting Node	Ending Node	Value of θ	Area	Modulus of Elasticity
1	1	2	90°	A	E
2	2	3	315°	1.5A	E
3	3	1	180°	A	E

Now, let assume the coordinate of node 1 as (0, 0). The coordinate and restraint joint information are given in the table (b). The integer 1 in the restraint list indicates the restraint exists and 0 indicates the restraint at that particular direction does not exist. Thus, in node no. 2, the integer 0 in x and y indicates that the joint is free in x and y directions.

(b) Nodal Information for Plane Truss

Node No.	Coordinates		Restraint List	
	x	y	x	y
1	O	O	1	1
2	O	L	O	O
3	L	O	1	1

The stiffness matrices of each individual member can be found out from the stiffness matrix equation as shown below:

$$[k] = \frac{AE}{L} \begin{bmatrix} \cos^2\theta & \cos\theta\sin\theta & -\cos^2\theta & -\cos\theta\sin\theta \\ \cos\theta\sin\theta & \sin^2\theta & -\cos\theta\sin\theta & -\sin^2\theta \\ -\cos^2\theta & -\cos\theta\sin\theta & \cos^2\theta & \cos\theta\sin\theta \\ -\cos\theta\sin\theta & -\sin^2\theta & \cos\theta\sin\theta & \sin^2\theta \end{bmatrix}$$

Thus, the local stiffness matrices of each member are calculated based on their individual member properties and orientations and written below:

$$[k]_1 = \frac{AE}{L} \begin{matrix} 1 & 2 & 3 & 4 \\ \begin{bmatrix} 0 & 0 & 0 & 0 \\ 0 & 1 & 0 & -1 \\ 0 & 0 & 0 & 0 \\ 0 & -1 & 0 & 1 \end{bmatrix} & \begin{matrix} 1 \\ 2 \\ 3 \\ 4 \end{matrix} \end{matrix}$$

$$[k]_2 = \frac{3AE}{4\sqrt{2}L} \begin{matrix} 3 & 4 & 5 & 6 \\ \begin{bmatrix} 1 & -1 & -1 & 1 \\ -1 & 1 & 1 & -1 \\ -1 & 1 & 1 & -1 \\ 1 & -1 & -1 & 1 \end{bmatrix} & \begin{matrix} 3 \\ 4 \\ 5 \\ 6 \end{matrix} \end{matrix}$$

And

$$[k]_3 = \frac{AE}{L} \begin{matrix} 5 & 6 & 1 & 2 \\ \begin{bmatrix} 1 & 0 & -1 & 0 \\ 0 & 0 & 0 & -1 \\ -1 & 0 & 1 & 0 \\ 0 & 0 & 0 & 0 \end{bmatrix} & \begin{matrix} 5 \\ 6 \\ 1 \\ 2 \end{matrix} \end{matrix}$$

Global stiffness matrix can be formed by assembling the local stiffness matrices into globally. Thus the global stiffness matrix are calculated from the above relations and obtained as follows:

$$
[K] = \frac{AE}{L}
\begin{array}{c}
\begin{array}{cccccc} 1 & 2 & 3 & 4 & 5 & 6 \end{array} \\
\begin{bmatrix}
1 & 0 & 0 & 0 & -1 & 0 \\
0 & 1 & 0 & -1 & 0 & 0 \\
0 & 0 & \dfrac{3}{4\sqrt{2}} & -\dfrac{3}{4\sqrt{2}} & -\dfrac{3}{4\sqrt{2}} & \dfrac{3}{4\sqrt{2}} \\
0 & -1 & -\dfrac{3}{4\sqrt{2}} & 1+\dfrac{3}{4\sqrt{2}} & \dfrac{3}{4\sqrt{2}} & -\dfrac{3}{4\sqrt{2}} \\
-1 & 0 & -\dfrac{3}{4\sqrt{2}} & \dfrac{3}{4\sqrt{2}} & \dfrac{3}{4\sqrt{2}}+1 & -\dfrac{3}{4\sqrt{2}} \\
0 & 0 & \dfrac{3}{4\sqrt{2}} & -\dfrac{3}{4\sqrt{2}} & -\dfrac{3}{4\sqrt{2}} & \dfrac{3}{4\sqrt{2}}
\end{bmatrix}
\begin{array}{c} 1 \\ 2 \\ 3 \\ 4 \\ 5 \\ 6 \end{array}
\end{array}
$$

The equivalent load vector for the given truss can be written as:

$$
\{F\} = \begin{Bmatrix} F_{x1} \\ F_{y1} \\ F_{x2} \\ F_{y2} \\ F_{x3} \\ F_{y3} \end{Bmatrix} = \begin{Bmatrix} 0 \\ 0 \\ 2P \\ P \\ 0 \\ 0 \end{Bmatrix}
$$

Let us assume that, u and v are the horizontal and vertical displacements respectively at joints. Thus, the displacement vector will be expressed as follows:

$$
\{d\} = \begin{Bmatrix} u_1 \\ v_1 \\ u_2 \\ v_2 \\ u_3 \\ v_3 \end{Bmatrix} = \begin{Bmatrix} 0 \\ 0 \\ u_2 \\ v_2 \\ 0 \\ 0 \end{Bmatrix}
$$

Therefore, the relationship between the force and the displacement will be:

$$\begin{Bmatrix} F_{x1} \\ F_{y1} \\ 2P \\ P \\ F_{x3} \\ F_{y3} \end{Bmatrix} = \frac{AE}{L} \begin{bmatrix} 1 & 0 & 0 & 0 & -1 & 0 \\ 0 & 1 & 0 & -1 & 0 & 0 \\ 0 & 0 & \dfrac{3}{4\sqrt{2}} & -\dfrac{3}{4\sqrt{2}} & -\dfrac{3}{4\sqrt{2}} & \dfrac{3}{4\sqrt{2}} \\ 0 & -1 & -\dfrac{3}{4\sqrt{2}} & 1+\dfrac{3}{4\sqrt{2}} & \dfrac{3}{4\sqrt{2}} & -\dfrac{3}{4\sqrt{2}} \\ -1 & 0 & -\dfrac{3}{4\sqrt{2}} & \dfrac{3}{4\sqrt{2}} & \dfrac{3}{4\sqrt{2}}+1 & -\dfrac{3}{4\sqrt{2}} \\ 0 & 0 & \dfrac{3}{4\sqrt{2}} & -\dfrac{3}{4\sqrt{2}} & -\dfrac{3}{4\sqrt{2}} & \dfrac{3}{4\sqrt{2}} \end{bmatrix} \begin{Bmatrix} 0 \\ 0 \\ u_2 \\ v_2 \\ 0 \\ 0 \end{Bmatrix}$$

From the above relation, the unknown displacements u2 and v2 can be found out through computer programming. However, as numbers of unknown displacements in this case are only two, the solution can be obtained by manual calculations.

The above equation may be rearranged with respect to unknown and known displacements in the following form:

$$\begin{Bmatrix} F_\alpha \\ F_\beta \end{Bmatrix} = \begin{bmatrix} k_{\alpha\alpha} & k_{\alpha\beta} \\ k_{\beta\alpha} & k_{\beta\beta} \end{bmatrix} \begin{Bmatrix} d_\alpha \\ d_\beta \end{Bmatrix}$$

Thus, the developed matrices for the truss problem can be rearranged as:

$$\begin{Bmatrix} 2P \\ P \\ F_{x1} \\ F_{x2} \\ F_{x3} \\ F_{y3} \end{Bmatrix} = \frac{AE}{L} \begin{bmatrix} \dfrac{3}{4\sqrt{2}} & \dfrac{-3}{4\sqrt{2}} & 0 & 0 & \dfrac{-3}{4\sqrt{2}} & \dfrac{3}{4\sqrt{2}} \\ \dfrac{-3}{4\sqrt{2}} & 1+\dfrac{3}{4\sqrt{2}} & 0 & -1 & \dfrac{3}{4\sqrt{2}} & \dfrac{-3}{4\sqrt{2}} \\ 0 & 0 & 1 & 0 & -1 & 0 \\ 0 & -1 & 0 & 1 & 0 & 0 \\ \dfrac{-3}{4\sqrt{2}} & \dfrac{3}{4\sqrt{2}} & -1 & 0 & \dfrac{3}{4\sqrt{2}}+1 & -\dfrac{3}{4\sqrt{2}} \\ \dfrac{3}{4\sqrt{2}} & \dfrac{-3}{4\sqrt{2}} & 0 & 0 & -\dfrac{3}{4\sqrt{2}} & \dfrac{3}{4\sqrt{2}} \end{bmatrix} \begin{Bmatrix} u_2 \\ v_2 \\ 0 \\ 0 \\ 0 \\ 0 \end{Bmatrix}$$

The above relation may be condensed into following:

$$\begin{Bmatrix} 2P \\ P \end{Bmatrix} = \frac{AE}{L} \begin{bmatrix} \dfrac{3}{4\sqrt{2}} & \dfrac{-3}{4\sqrt{2}} \\ \dfrac{-3}{4\sqrt{2}} & 1+\dfrac{3}{4\sqrt{2}} \end{bmatrix} \begin{Bmatrix} u_2 \\ v_2 \end{Bmatrix}$$

The unknown displacements can be derived from the relationships expressed in the above equation:

$$\left\{\begin{matrix} u_2 \\ v_2 \end{matrix}\right\} = \frac{AE}{L}\begin{bmatrix} \dfrac{3}{4\sqrt{2}} & \dfrac{-3}{4\sqrt{2}} \\ \dfrac{-3}{4\sqrt{2}} & 1+\dfrac{3}{4\sqrt{2}} \end{bmatrix}\left\{\begin{matrix} 2P \\ P \end{matrix}\right\} = \frac{4\sqrt{2}\,L}{3AE}\begin{bmatrix} 1+\dfrac{3}{4\sqrt{2}} & \dfrac{3}{4\sqrt{2}} \\ \dfrac{3}{4\sqrt{2}} & \dfrac{3}{4\sqrt{2}} \end{bmatrix}\left\{\begin{matrix} 2P \\ P \end{matrix}\right\}$$

Thus, the unknown displacement at node 2 of the truss structure will become:

$$\left\{\begin{matrix} u_2 \\ v_2 \end{matrix}\right\} = \frac{PL}{AE}\begin{bmatrix} 3+\dfrac{8\sqrt{2}}{3} \\ 3 \end{bmatrix}$$

Support Reactions:

The support reactions $\{P_s\}$ can be determined from the following relation:

$$\{P_s\} = -\{P_{cs}\} + [K_{\beta\alpha}]\{d_\alpha\}$$

Where, $\{P_{cs}\}$ correspond to equivalent loadings at supports. Thus, the support reaction of the present truss structure will be:

$$\{P_s\} = \left\{\begin{matrix} 0 \\ 0 \\ 0 \\ 0 \end{matrix}\right\} + \frac{AE}{L}\begin{bmatrix} 0 & 0 \\ 0 & -1 \\ \dfrac{-3}{4\sqrt{2}} & \dfrac{3}{4\sqrt{2}} \\ \dfrac{3}{4\sqrt{2}} & \dfrac{-3}{4\sqrt{2}} \end{bmatrix} \frac{PL}{AE}\begin{bmatrix} 3+\dfrac{8\sqrt{2}}{3} \\ 3 \end{bmatrix} = \begin{bmatrix} 0 \\ -3P \\ -2P \\ 2P \end{bmatrix}$$

Member End Actions

Now, the member end actions can be obtained from the corresponding member stiffness and the nodal displacements. The member end forces are derived as shown below:

Member -1:

$$\left\{\begin{matrix} F_{mx1} \\ F_{my1} \\ F_{mx2} \\ F_{my2} \end{matrix}\right\} = \frac{AE}{L}\begin{bmatrix} 0 & 0 & 0 & 0 \\ 0 & 1 & 0 & -1 \\ 0 & 0 & 0 & 0 \\ 0 & -1 & 0 & 1 \end{bmatrix}\left\{\begin{matrix} 0 \\ 0 \\ 3+\dfrac{8\sqrt{2}}{3} \\ 3 \end{matrix}\right\}\frac{PL}{AE} = \left\{\begin{matrix} 0 \\ -3P \\ 0 \\ 3P \end{matrix}\right\}$$

Member -2:

$$
\begin{Bmatrix} F_{mx1} \\ F_{my1} \\ F_{mx2} \\ F_{my2} \end{Bmatrix} = \frac{AE}{4\sqrt{2}L} \begin{bmatrix} 1 & -1 & -1 & 1 \\ -1 & 1 & 1 & -1 \\ -1 & 1 & 1 & -1 \\ 1 & -1 & -1 & 1 \end{bmatrix} \begin{Bmatrix} 3+\dfrac{8\sqrt{2}}{3} \\ 3 \\ 0 \\ 0 \end{Bmatrix} \frac{PL}{AE} = \begin{Bmatrix} 2P \\ -2P \\ -2P \\ 2P \end{Bmatrix}
$$

Member -3:

$$
\begin{Bmatrix} F_{mx3} \\ F_{my3} \\ F_{mx1} \\ F_{my1} \end{Bmatrix} = \frac{AE}{L} \begin{bmatrix} 1 & 0 & -1 & 0 \\ 0 & 0 & 0 & 0 \\ -1 & 0 & 1 & 0 \\ 0 & 0 & 0 & 0 \end{bmatrix} \begin{Bmatrix} 0 \\ 0 \\ 0 \\ 0 \end{Bmatrix} \frac{PL}{AE} = \begin{Bmatrix} 0 \\ 0 \\ 0 \\ 0 \end{Bmatrix}
$$

Thus, the member forces in all members of the truss will be:

$$
\{F_m\} = \begin{Bmatrix} 3P \\ -\sqrt{(2P)^2 + (2P)^2} \\ 0 \end{Bmatrix} = \begin{Bmatrix} 3P \\ -2\sqrt{2}P \\ 0 \end{Bmatrix}
$$

The reaction forces at the supports of the truss structure will be:

$$
\{F_R\} = \begin{Bmatrix} 0 \\ -3P \\ -2P \\ 2P \end{Bmatrix}
$$

Thus, the member force diagram will be as shown in the figure (c):

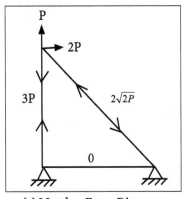

(c) Member Force Diagram.

4.3   Stiffness of Beam Members

A beam is a structural member which is capable of withstanding load primarily by resisting bending. The primary tool for analysis of beam is the Euler–Bernoulli beam equation. Other methods for determining the deflection of beams include "slope deflection method" and "method of virtual work".

For calculation of internal forces of beam include "moment distribution method", force or flexibility method and stiffness method. However, all these methods have limitations if either of geometry, loading, material properties or boundary conditions becomes arbitrary in nature.

Finite elements techniques can well handle such cases and relieve the analyzer of making simplifications to arrive approximate solutions.

Derivation of Shape Function

The degrees of freedom at each node for a beam member will be (i) vertical deflection and (ii) rotation. For a beam member, the slope of the elastic curve θ is given by, $\theta = \dfrac{dv}{dx}$, where the variable v is the displacement function of the beam. As the beam has two degrees of freedom at each node, the variation of v will be cubic and can be expressed using Pascal's triangle as:

$$v(x) = \alpha_0 + \alpha_1 x + \alpha_2 x^2 + \alpha_3 x^3 = \begin{bmatrix} 1 & x & x^2 & x^3 \end{bmatrix} \begin{Bmatrix} \alpha_0 \\ \alpha_1 \\ \alpha_2 \\ \alpha_3 \end{Bmatrix} \qquad ...(1)$$

And

$$\theta = \frac{dv}{dx} = \begin{bmatrix} 0 & 1 & 2x & 3x^2 \end{bmatrix} \begin{Bmatrix} \alpha_0 \\ \alpha_1 \\ \alpha_2 \\ \alpha_3 \end{Bmatrix} \qquad ...(2)$$

(a) Beam element.

Now, applying boundary conditions, the following expressions from the above relations can be obtained:

At x=0

$$v_1 = \begin{bmatrix} 1 & 0 & 0 & 0 \end{bmatrix} \begin{Bmatrix} \alpha_0 \\ \alpha_1 \\ \alpha_2 \\ \alpha_3 \end{Bmatrix} ; \ \theta_1 = \begin{bmatrix} 0 & 1 & 0 & 0 \end{bmatrix} \begin{Bmatrix} \alpha_0 \\ \alpha_1 \\ \alpha_2 \\ \alpha_3 \end{Bmatrix}$$

At x=L:

$$v_2 = \begin{bmatrix} 1 & L & L^2 & L^3 \end{bmatrix} \begin{Bmatrix} \alpha_0 \\ \alpha_1 \\ \alpha_2 \\ \alpha_3 \end{Bmatrix} ; \ \theta_2 = \begin{bmatrix} 0 & 1 & 2L & 3L^2 \end{bmatrix} \begin{Bmatrix} \alpha_0 \\ \alpha_1 \\ \alpha_2 \\ \alpha_3 \end{Bmatrix}$$

Thus, combining the above expressions one can write:

$$\begin{Bmatrix} V_1 \\ \theta_1 \\ V_2 \\ \theta_2 \end{Bmatrix} = \begin{bmatrix} 1 & 0 & 0 & 0 \\ 0 & 1 & 0 & 0 \\ 1 & L & L^2 & L^3 \\ 0 & 1 & 2L & 3L^2 \end{bmatrix} \begin{Bmatrix} \alpha_0 \\ \alpha_1 \\ \alpha_2 \\ \alpha_3 \end{Bmatrix} = [A]\{\alpha\} \qquad ....(3)$$

So,

$$\begin{Bmatrix} \alpha_0 \\ \alpha_1 \\ \alpha_2 \\ \alpha_3 \end{Bmatrix} = \begin{bmatrix} 1 & 0 & 0 & 0 \\ 0 & 1 & 0 & 0 \\ 1 & L & L^2 & L^3 \\ 0 & 1 & 2L & 3L^2 \end{bmatrix}^{-1} \begin{Bmatrix} V_1 \\ \theta_1 \\ V_2 \\ \theta_2 \end{Bmatrix} = \begin{bmatrix} 1 & 0 & 0 & 0 \\ 0 & 1 & 0 & 0 \\ -\dfrac{3}{L^2} & -\dfrac{2}{L} & \dfrac{3}{L^2} & -\dfrac{1}{L} \\ \dfrac{2}{L^3} & \dfrac{1}{L^2} & -\dfrac{2}{L^3} & \dfrac{1}{L^2} \end{bmatrix} \begin{Bmatrix} V_1 \\ \theta_1 \\ V_2 \\ \theta_2 \end{Bmatrix} \qquad ...(4)$$

Therefore,

$$v(x) = \begin{bmatrix} 1 & x & x^2 & x^3 \end{bmatrix} \begin{bmatrix} 1 & 0 & 0 & 0 \\ 0 & 1 & 0 & 0 \\ -\dfrac{3}{l^2} & -\dfrac{2}{l} & \dfrac{3}{l^2} & -\dfrac{1}{l} \\ \dfrac{2}{l^3} & \dfrac{1}{l^2} & -\dfrac{2}{l^3} & \dfrac{1}{l^2} \end{bmatrix} \begin{Bmatrix} V_1 \\ \theta_1 \\ V_2 \\ \theta_2 \end{Bmatrix} = \begin{bmatrix} N_1 & N_2 & N_3 & N_4 \end{bmatrix} \begin{Bmatrix} V_1 \\ \theta_1 \\ V_2 \\ \theta_2 \end{Bmatrix} \ ...(5)$$

Where,

$$N_1 = 1 - \frac{3}{L^2}x^2 + \frac{2}{L^3}x^3; \quad N_2 = x - \frac{2}{L}x^2 + \frac{x^3}{L^2}; N_3 = \frac{3x^2}{L^2} - \frac{2x^3}{L^3} \text{ and } N_4 = -\frac{x^2}{L} + \frac{x^3}{L^2} \qquad ...(6)$$

N is called shape function which interpolates the beam displacement in terms of its nodal displacements.

Derivation of Element Stiffness Matrix

Now, the strain displacement relationship matrix [B] can be expressed from the following expressions with the help of equation (1):

$$\chi = \frac{d^2v}{dx^2} = \begin{bmatrix} 0 & 0 & 2 & 6x \end{bmatrix} \begin{Bmatrix} \alpha_0 \\ \alpha_1 \\ \alpha_2 \\ \alpha_3 \end{Bmatrix} = [B]\{\alpha\} = [B][A]^{-1}\{d\} \qquad ...(7)$$

Where, $[B] = \begin{bmatrix} 0 & 0 & 2 & 6x \end{bmatrix}; [A] = \begin{bmatrix} 1 & 0 & 0 & 0 \\ 0 & 1 & 0 & 0 \\ 1 & L & L^2 & L^3 \\ 0 & 1 & 2L & 3L^2 \end{bmatrix}; \{d\} = \begin{Bmatrix} V_1 \\ \theta_1 \\ V_2 \\ \theta_2 \end{Bmatrix}$

From the moment curvature relationship, we can write:

$$M = EI\chi = EI\frac{d^2v}{dx^2} = EI[B][A]^{-1}\{d\} \qquad ...(8)$$

Strain energy:

$$U = \int_0^L \frac{1}{2}[\chi]^T[M]dx = \frac{EI}{2}\int_0^L \{d\}^T [A^{-1}]^T [B]^T [B][A^{-1}]\{d\}dx \qquad ...(9)$$

Thus:

$$\{F\} = \frac{\partial U}{\partial\{d\}} = EI\int_0^L [A^{-1}]^T [B]^T [B][A^{-1}]\{d\}dx \qquad ...(10)$$

$$[k] = EI\int_0^L [A^{-1}]^T [B]^T [B][A^{-1}]dx = EI[A^{-1}]^T \int_0^L [B]^T [B]dx[A]^{-1} \qquad ...(11)$$

Now,

$$\int_0^L \left[B^T\right]\left[B\right] dx = \int_0^L \begin{Bmatrix} 0 \\ 0 \\ 2 \\ 6x \end{Bmatrix} \begin{bmatrix} 0 & 0 & 2 & 6x \end{bmatrix} dx = \int_0^L \begin{bmatrix} 0 & 0 & 0 & 0 \\ 0 & 0 & 0 & 0 \\ 0 & 0 & 4 & 12x \\ 0 & 0 & 12x & 36x^2 \end{bmatrix}$$

$$dx = \int_0^L \begin{bmatrix} 0 & 0 & 0 & 0 \\ 0 & 0 & 0 & 0 \\ 0 & 0 & 4 & 12x \\ 0 & 0 & 12x & 36x^2 \end{bmatrix} dx = \begin{bmatrix} 0 & 0 & 0 & 0 \\ 0 & 0 & 0 & 0 \\ 0 & 0 & 4L & 6L^2 \\ 0 & 0 & 6L^2 & 12L^3 \end{bmatrix} \qquad ...(12)$$

So,

$$[k] = EI\left[A^{-1}\right]^T \begin{bmatrix} 0 & 0 & 0 & 0 \\ 0 & 0 & 0 & 0 \\ 0 & 0 & 4L & 6L^2 \\ 0 & 0 & 6L^2 & 12L^3 \end{bmatrix} \left[A\right]^{-1}$$

$$[k] = EI \begin{bmatrix} 1 & 0 & -\dfrac{3}{L^2} & \dfrac{2}{L^3} \\ 0 & 1 & -\dfrac{2}{L} & \dfrac{1}{L^2} \\ 0 & 0 & \dfrac{3}{L^2} & -\dfrac{2}{L^3} \\ 0 & 0 & -\dfrac{1}{L} & \dfrac{1}{L^2} \end{bmatrix} \begin{bmatrix} 0 & 0 & 0 & 0 \\ 0 & 0 & 0 & 0 \\ 0 & 0 & 4L & 6L^2 \\ 0 & 0 & 6L^2 & 12L^3 \end{bmatrix} \begin{bmatrix} 1 & 0 & 0 & 0 \\ 0 & 1 & 0 & 0 \\ -\dfrac{3}{L^2} & \dfrac{2}{L} & \dfrac{3}{L^2} & -\dfrac{1}{L} \\ \dfrac{2}{L^3} & \dfrac{1}{L^2} & -\dfrac{2}{L^3} & \dfrac{1}{L^2} \end{bmatrix}$$

$$= EI \begin{bmatrix} 0 & 0 & 0 & 0 \\ 0 & 0 & -2 & 0 \\ 0 & 0 & 0 & -6 \\ 0 & 0 & 2 & 6l \end{bmatrix} \begin{bmatrix} 1 & 0 & 0 & 0 \\ 0 & 1 & 0 & 0 \\ -\dfrac{3}{L^2} & -\dfrac{2}{L} & \dfrac{3}{L^2} & -\dfrac{1}{L} \\ \dfrac{2}{L^3} & \dfrac{1}{L^2} & -\dfrac{2}{L^3} & \dfrac{1}{L^2} \end{bmatrix} = EI \begin{bmatrix} \dfrac{12}{L^3} & \dfrac{6}{L^2} & \dfrac{12}{L^3} & \dfrac{6}{L^2} \\ \dfrac{6}{L^2} & \dfrac{4}{L} & -\dfrac{6}{L^2} & \dfrac{2}{L} \\ -\dfrac{12}{L^2} & -\dfrac{6}{L^2} & \dfrac{12}{L^3} & \dfrac{6}{L^2} \\ \dfrac{6}{L^2} & \dfrac{2}{L} & -\dfrac{6}{L^2} & \dfrac{4}{L} \end{bmatrix}$$

Thus, the element stiffness of a beam member is:

$$[k] = \dfrac{EI}{L^3} \begin{bmatrix} 12 & 6L & -12 & 6L \\ 6L & 4L^2 & -6L & 2L^2 \\ -12 & -6L & 12 & -6L \\ 6L & 2L^2 & -6L & 4L^2 \end{bmatrix} \qquad ...(13)$$

Generalized Stiffness Matrix of a Beam Member

Consider a beam member making an angle 'θ' with X axis as shown in the figure (b) below. By resolving the forces along local X and Y direction, the following relations are obtained:

$$
\left.\begin{aligned}
\overline{F}_{x1} &= F_{x1}\cos\theta + F_{y1}\sin\theta \\
\overline{F}_{x2} &= F_{x2}\cos\theta + F_{y2}\sin\theta \\
\overline{F}_{y2} &= -F_{x1}\sin\theta + F_{y1}\cos\theta \\
\overline{F}_{y2} &= -F_{x2}\sin\theta + F_{y2}\cos\theta \\
\overline{M}_1 &= M_1 \\
\overline{M}_2 &= M_2
\end{aligned}\right\} \qquad ...(14)
$$

Where, $\overline{F}_{x1}$ and $\overline{F}_{x2}$ are the axial forces along the member $\overline{X}$ axis. Similarly, $\overline{F}_{y1}$ and $\overline{F}_{y2}$ are the forces perpendicular to the member axis $\overline{X}$. $\overline{M}_1$ and $\overline{M}_2$ are the moment about its axis at node 1 and 2 respectively.

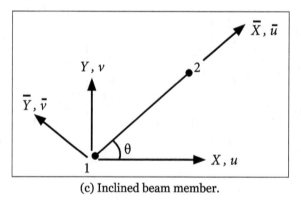

(c) Inclined beam member.

The relationship expressed in equation (14) can be rewritten in matrix form as follows:

$$
\begin{Bmatrix}
\overline{F}_{x1} \\
\overline{F}_{y1} \\
\overline{M}_1 \\
\overline{F}_{x2} \\
\overline{F}_{y2} \\
\overline{M}_2
\end{Bmatrix}
=
\begin{bmatrix}
\cos\theta & \sin\theta & 0 & 0 & 0 & 0 \\
-\sin\theta & \cos\theta & 0 & 0 & 0 & 0 \\
0 & 0 & 1 & 0 & 0 & 0 \\
0 & 0 & 0 & \cos\theta & \sin\theta & 0 \\
0 & 0 & 0 & -\sin\theta & \cos\theta & 0 \\
0 & 0 & 0 & 0 & 0 & 1
\end{bmatrix}
\begin{Bmatrix}
F_{x1} \\
F_{y1} \\
M_1 \\
F_{x2} \\
F_{y2} \\
M_2
\end{Bmatrix}
\qquad ...(15)
$$

Now, the above equation can be expressed in short as:

$$\{\overline{F}\}=[T]\{F\} \qquad ...(16)$$

Similarly, the displacement vector in local coordinate system $(\overline{X}, \overline{Y})$ may be transformed to global (X,Y) coordinate system by the following relation:

$$\{\overline{d}\} = [T]\{d\} \qquad \qquad ...(17)$$

The force-displacement relation in local coordinate system may be expressed as:

$$
\begin{Bmatrix} \overline{F}_{x1} \\ \overline{F}_{y1} \\ \overline{M}_1 \\ \overline{F}_{x2} \\ \overline{F}_{y2} \\ \overline{M}_2 \end{Bmatrix} =
\begin{bmatrix}
0 & 0 & 0 & 0 & 0 & 0 \\
0 & \dfrac{12\,EI}{L^3} & \dfrac{6\,EI}{L^2} & 0 & -\dfrac{12\,EI}{L^3} & \dfrac{6\,EI}{L^2} \\
0 & \dfrac{6\,EI}{L^2} & \dfrac{4\,EI}{L} & 0 & -\dfrac{6\,EI}{L^2} & \dfrac{2\,EI}{L} \\
0 & 0 & 0 & 0 & 0 & 0 \\
0 & -\dfrac{12\,EI}{L^3} & -\dfrac{6\,EI}{L^2} & 0 & \dfrac{12\,EI}{L^3} & -\dfrac{6\,EI}{L^2} \\
0 & \dfrac{6\,EI}{L^2} & \dfrac{2\,EI}{L} & 0 & -\dfrac{6\,EI}{L^2} & \dfrac{4\,EI}{L}
\end{bmatrix}
\begin{Bmatrix} \overline{u}_1 \\ \overline{v}_1 \\ \overline{\theta}_1 \\ \overline{u}_2 \\ \overline{v}_2 \\ \overline{\theta}_2 \end{Bmatrix} \qquad ...(18)
$$

The matrices in the above equation are written with respect to the member axis. Now, the equation (18) can be rewritten as follows with the use of equations (16) and (17).

$$[T]\{F\} = [\overline{k}][T]\{d\} \qquad \qquad ...(19)$$

Or

$$\{F\} = [T]^{-1}[\overline{k}][T]\{d\} \qquad \qquad ...(20)$$

Here, the transformation matrix [T] is orthogonal. Thus, from the above relationship, the generalized stiffness matrix can be expressed as:

$$[k] = \left[[T]^{T}[\overline{k}]\right][T] \qquad \qquad ...(21)$$

Considering $\lambda = \cos\theta$ and $\mu = \sin\theta$ the above expression can be written as follows:

$$
[k] = EI
\begin{bmatrix}
\lambda & -\mu & 0 & 0 & 0 & 0 \\
\mu & \lambda & 0 & 0 & 0 & 0 \\
0 & 0 & 1 & 0 & 0 & 0 \\
0 & 0 & 0 & \lambda & -\mu & 0 \\
0 & 0 & 0 & \mu & \lambda & 0 \\
0 & 0 & 0 & 0 & 0 & 0
\end{bmatrix}
\begin{bmatrix}
0 & 0 & 0 & 0 & 0 & 0 \\
0 & \dfrac{12}{L^3} & \dfrac{6}{L^2} & 0 & -\dfrac{12}{L^3} & \dfrac{6}{L^2} \\
0 & \dfrac{6}{L^2} & \dfrac{4}{L} & 0 & -\dfrac{6}{L^2} & \dfrac{2}{L} \\
0 & 0 & 0 & 0 & 0 & 0 \\
0 & -\dfrac{12}{L^3} & -\dfrac{6}{L^2} & 0 & \dfrac{12}{L^3} & -\dfrac{6}{L^2} \\
0 & \dfrac{6}{L^2} & \dfrac{2}{L} & 0 & -\dfrac{6}{L^2} & \dfrac{4}{L}
\end{bmatrix}
$$

$$
\begin{vmatrix}
\lambda & \mu & 0 & 0 & 0 & 0 \\
-\mu & \lambda & 0 & 0 & 0 & 0 \\
0 & 0 & 1 & 0 & 0 & 0 \\
0 & 0 & 0 & \lambda & \mu & 0 \\
0 & 0 & 0 & -\mu & \lambda & 0 \\
0 & 0 & 0 & 0 & 0 & 1
\end{vmatrix}
\qquad \dots(22)
$$

Thus, the generalized stiffness matrix of a beam member is derived as:

$$
[k] = EI
\begin{bmatrix}
\dfrac{12\mu^2}{L^3} & -\dfrac{12\mu\lambda}{L^3} & -\dfrac{6\mu^2}{L^2} & -\dfrac{12\mu^2}{L^3} & \dfrac{12\mu\lambda}{L^3} & -\dfrac{6\mu}{L^2} \\[2mm]
-\dfrac{12\mu\lambda}{L^3} & \dfrac{12\lambda^2}{L^3} & \dfrac{6\lambda}{L^2} & \dfrac{12\mu\lambda}{L^3} & -\dfrac{12\lambda^2}{L^3} & \dfrac{6\lambda}{L^2} \\[2mm]
-\dfrac{6\mu}{L^2} & \dfrac{6\lambda}{L^2} & \dfrac{4}{L} & \dfrac{6\mu}{L^2} & -\dfrac{6\lambda}{L^2} & \dfrac{2}{L} \\[2mm]
-\dfrac{12\mu^2}{L^3} & \dfrac{12\mu\lambda}{L^3} & \dfrac{6\mu}{L^2} & \dfrac{12\mu^2}{L^3} & -\dfrac{12\mu\lambda}{L^3} & \dfrac{6\mu}{L^2} \\[2mm]
\dfrac{12\mu\lambda}{L^3} & -\dfrac{12\lambda^2}{L^3} & -\dfrac{6\lambda}{L^2} & -\dfrac{12\mu\lambda}{L^3} & \dfrac{12\lambda^2}{L^3} & \dfrac{6\lambda}{L^2} \\[2mm]
-\dfrac{6\mu}{L^2} & \dfrac{6\lambda}{L^2} & \dfrac{2}{L} & \dfrac{6\mu}{L^2} & -\dfrac{6\lambda}{L^2} & \dfrac{4}{L}
\end{bmatrix}
\qquad \dots(23)
$$

4.4 Finite Element Analysis of Continuous Beam

Equivalent Loading on Beam Member

In finite element analysis, the external loads are necessary to be acting at the joints, which does not happen always; as some forces may act on the member. The forces acting on the member should be replaced by equivalent forces acting at the joints.

These joint forces obtained from the forces on the members are called equivalent joint loads. These joint loads are combined with the actual joint loads to provide the combined joint loads, which are then utilized in the analysis.

Varying Load

Let a beam is loaded with a linearly varying load as shown in the figure below. The equivalent forces at nodes can be expressed using finite element technique. If w(x) is the function of load, then the nodal load can be expressed as follows.

$$
\{Q\} = \int [N]^T w(x)\, dx
\qquad \dots(1)
$$

The loading function for the present case can be written as:

$$w(x) = w_1 + \frac{w_2 - w_1}{L}x \qquad \qquad ...(2)$$

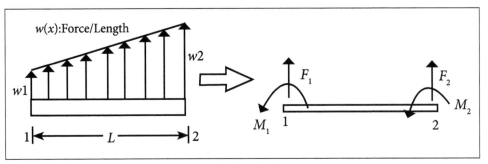

(a) Varying load on beam:

$$\{Q\} = \begin{Bmatrix} F_1 \\ M_1 \\ F_2 \\ M_2 \end{Bmatrix} = \begin{Bmatrix} \int_0^L \left(\frac{2x^3}{L^3} - \frac{3x^2}{L^2} + 1 \right) w(x)dx \\ \int_0^L \left(\frac{x^3}{L^2} - \frac{2x^2}{L} + x \right) w(x)dx \\ \int_0^L \left(-\frac{2x^3}{L^3} + \frac{3x^2}{L^2} \right) w(x)dx \\ \int_0^L \left(\frac{x^3}{L^2} - \frac{x^2}{L} \right) w(x)dx \end{Bmatrix} = \begin{Bmatrix} \left(\frac{7w_1}{20} + \frac{3w_2}{20} \right)L \\ \left(\frac{w_1}{20} + \frac{w_2}{30} \right)L^2 \\ \left(\frac{3w_1}{20} + \frac{7w_2}{20} \right)L \\ \left(-\frac{w_1}{30} - \frac{w_2}{20} \right)L^2 \end{Bmatrix} \qquad ...(3)$$

Now, if $w_1 = w_2 = w$, then the equivalent nodal force will be:

$$\{Q\} = \begin{Bmatrix} \dfrac{wL}{2} \\ \dfrac{wL^2}{12} \\ \dfrac{wL}{2} \\ -\dfrac{wL^2}{12} \end{Bmatrix} \qquad \qquad ...(4)$$

Concentrated Load

Consider a force F is applied at a point is regarded as a limiting case of intense pressure over infinitesimal length, so that p(x) dx approaches F. Therefore:

$$\{Q\} = \int \{N\}^T p(x)dx = \{N^\cdot\}^T F \qquad \qquad ...(5)$$

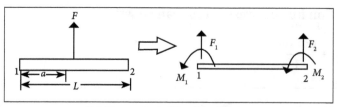

(b) Concentrated load on beam.

Here, [N*] is obtained by evaluating [N] at point where the concentrated load F is applied. Thus:

$$\{N*\}^T = \begin{Bmatrix} \left(\dfrac{2x^3}{L^3} - \dfrac{3x^2}{L^2} + 1\right) \\[2mm] \left(\dfrac{x^3}{L^2} - \dfrac{2x^2}{L} + x\right) \\[2mm] \left(-\dfrac{2x^3}{L^3} + \dfrac{3x^2}{L^2}\right) \\[2mm] \left(\dfrac{x^3}{L^2} - \dfrac{x^2}{L}\right) \end{Bmatrix} \text{ at distance a} = \begin{Bmatrix} \dfrac{2a^3}{L^3} - \dfrac{3a^2}{L^2} + 1 \\[2mm] \dfrac{a^3}{L^2} - \dfrac{2a^2}{L} + a \\[2mm] -\dfrac{2a^3}{L^3} + \dfrac{3a^2}{L^2} \\[2mm] \dfrac{a^3}{L^3} - \dfrac{a^2}{L} \end{Bmatrix} \qquad ...(6)$$

Therefore,

$$\{Q\} = \begin{Bmatrix} F_1 \\ M_1 \\ F_2 \\ M_2 \end{Bmatrix} = \begin{Bmatrix} \left(\dfrac{2a^3}{L^3} - \dfrac{3a^2}{L^2} + 1\right)F \\[2mm] \left(\dfrac{a^3}{L^2} - \dfrac{2a^2}{L} + a\right)F \\[2mm] \left(-\dfrac{2a^3}{L^3} + \dfrac{3a^2}{L^2}\right)F \\[2mm] \left(\dfrac{a^3}{L^2} - \dfrac{a^2}{L}\right)F \end{Bmatrix} \qquad ...(7)$$

Now, if load F is acting at mid-span (i.e., a=L/2), then equivalent nodal load will be:

$$\{Q\} = \begin{Bmatrix} \dfrac{F}{2} \\[2mm] \dfrac{FL}{8} \\[2mm] \dfrac{F}{2} \\[2mm] -\dfrac{FL}{8} \end{Bmatrix} \qquad ...(8)$$

With the above approach, the equivalent nodal load can be found for various loading function acting on beam members.

Problem:

1. Let us analyze the beam shown below by the finite element method. Assume the moment of inertia of member 2 as twice that of member 1. Let us find the bending moment and reactions at supports of the beam assuming the length of span, L as 4 m, concentrated load (P) as 15 kN and udl, w as 4 kN/m.

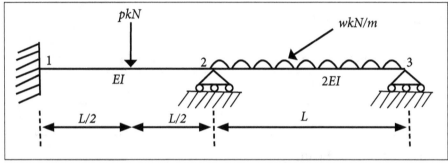

(a) Example of a continuous beam.

Solution

Given:

Assume the moment of inertia of member 2 as twice that of member 1.

Assume the length of span, L as 4 m, concentrated load (P) as 15 kN and udl, w as 4 kN/m.

Step 1: Numbering of Nodes and Members

The analysis of beam starts with the numbering of members and joints as shown below:

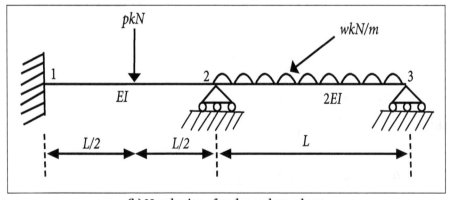

(b) Numbering of nodes and members.

The member AB and BC are designated as (1) and (2). The points A, B, C are designated by nodes 1, 2 and 4. The member information for beam is shown in tabulated form as shown in the table (a).

The coordinate of node 1 is assumed as (0, 0). The coordinate and restraint joint information are shown in the table (b). The integer 1 in the restraint list indicates the restraint exists and 0 indicates the restraint at that particular direction does not exist. Thus, in node no. 2, the integer 0 in rotation indicates that the joint is free rotation.

(a)Member Information for Beam

Member number	Starting node	Ending node	Rigidity modulus
1	1	2	EI
2	2	3	2EI

(b) Nodal Information for Beam

Node No.	Coordinates		Restraint List	
	x	y	Vertical	Rotation
1	0	0	1	1
2	L	0	1	0
3	2L	0	1	0

Step 2: Formation of Member Stiffness Matrix

The local stiffness matrices of each member are given below based on their individual member properties and orientations. Thus the local stiffness matrix of member (1) is given by:

$$
[k]_1 =
\begin{array}{cccc}
 1 & 2 & 3 & 4 \\
\end{array}
\begin{bmatrix}
\dfrac{12\,EI}{L^3} & \dfrac{6\,EI}{L^2} & -\dfrac{12\,EI}{L^3} & \dfrac{6\,EI}{L^2} \\[2ex]
\dfrac{6\,EI}{L^2} & \dfrac{4\,EI}{L} & -\dfrac{6\,EI}{L^2} & \dfrac{2\,EI}{L} \\[2ex]
-\dfrac{12\,EI}{L^3} & -\dfrac{6\,EI}{L^2} & \dfrac{12\,EI}{L^3} & -\dfrac{6\,EI}{L^2} \\[2ex]
\dfrac{6\,EI}{L^2} & \dfrac{2\,EI}{L} & -\dfrac{6\,EI}{L^2} & \dfrac{4\,EI}{L}
\end{bmatrix}
\begin{array}{c}
1 \\[2ex] 2 \\[2ex] 3 \\[2ex] 4
\end{array}
$$

Similarly, the local stiffness matrix of member (2) is given by:

$$
[k]_2 =
\begin{array}{c c c c c}
 & 3 & 4 & 5 & 6 \\
\end{array}
\left[
\begin{array}{cccc}
\dfrac{24\,EI}{L^3} & \dfrac{12\,EI}{L^2} & -\dfrac{24\,EI}{L^3} & \dfrac{12\,EI}{L^2} \\[2ex]
\dfrac{12\,EI}{L^2} & \dfrac{8\,EI}{L} & -\dfrac{12\,EI}{L^2} & \dfrac{4\,EI}{L} \\[2ex]
-\dfrac{24\,EI}{L^3} & -\dfrac{12\,EI}{L^2} & \dfrac{24\,EI}{L^3} & -\dfrac{12\,EI}{L^2} \\[2ex]
\dfrac{12\,EI}{L^2} & \dfrac{4\,EI}{L} & -\dfrac{12\,EI}{L^2} & \dfrac{8\,EI}{L}
\end{array}
\right]
\begin{array}{c}
3 \\[2ex] 4 \\[2ex] 5 \\[2ex] 6
\end{array}
$$

Step 3: Formation of Global Stiffness Matrix

The global stiffness matrix is obtained by assembling the local stiffness matrix of members (1) and (2) as follows:

$$
[K] =
\begin{array}{c c c c c c}
1 & 2 & 3 & 4 & 5 & 6 \\
\end{array}
\left[
\begin{array}{cccccc}
\dfrac{12\,EI}{L^3} & \dfrac{6\,EI}{L^2} & -\dfrac{12\,EI}{L^3} & \dfrac{6\,EI}{L^2} & 0 & 0 \\[2ex]
\dfrac{6\,EI}{L^2} & \dfrac{4\,EI}{L} & -\dfrac{6\,EI}{L^2} & \dfrac{2\,EI}{L} & 0 & 0 \\[2ex]
-\dfrac{12\,EI}{L^3} & -\dfrac{6\,EI}{L^2} & \dfrac{36\,EI}{L} & \dfrac{6\,EI}{L^2} & -\dfrac{24\,EI}{L^3} & \dfrac{12\,EI}{L^2} \\[2ex]
\dfrac{6\,EI}{L^2} & \dfrac{2\,EI}{L} & \dfrac{6\,EI}{L^2} & \dfrac{12\,EI}{L} & \dfrac{12\,EI}{L^2} & \dfrac{4\,EI}{L} \\[2ex]
0 & 0 & -\dfrac{24\,EI}{L^3} & -\dfrac{12\,EI}{L^2} & \dfrac{24\,EI}{L^3} & -\dfrac{12\,EI}{L^2} \\[2ex]
0 & 0 & \dfrac{12\,EI}{L^2} & \dfrac{4\,EI}{L} & -\dfrac{12\,EI}{L^2} & \dfrac{8\,EI}{L}
\end{array}
\right]
\begin{array}{c}
1 \\[2ex] 2 \\[2ex] 3 \\[2ex] 4 \\[2ex] 5 \\[2ex] 6
\end{array}
$$

Step 4: Boundary Condition

The boundary conditions according to the support of the beam can be expressed in terms of the displacement vector. The displacement vector will be as follows:

$$
\{d\} =
\begin{Bmatrix}
0 \\
0 \\
0 \\
\theta_2 \\
0 \\
\theta_3
\end{Bmatrix}
$$

Step 5: Load Vector

The concentrated load on member (1) and the distributed load on member (2) are replaced by equivalent joint load. The equivalent joint load vector can be written as:

$$\{F\}=\begin{Bmatrix} -\dfrac{P}{2} \\[2mm] -\dfrac{PL}{8} \\[2mm] -\left(\dfrac{P}{2}+\dfrac{wL}{2}\right) \\[2mm] \left(\dfrac{PL}{8}-\dfrac{wL^2}{12}\right) \\[2mm] -\dfrac{wL}{2} \\[2mm] \dfrac{wL^2}{12} \end{Bmatrix}$$

(c) Equivalent Load

Step 6: Determination of Unknown Displacements

The unknown displacement can be obtained from the relationship as given below:

$$\{F\}=[K]\{d\}$$

$$\{d\}=[K]^{-1}\{F\}$$

$$\begin{Bmatrix} 0 \\ 0 \\ 0 \\ 0 \\ \theta_2 \\ 0 \\ \theta_3 \end{Bmatrix}=\begin{bmatrix} \dfrac{12EI}{L^3} & \dfrac{6EI}{L^2} & -\dfrac{12EI}{L^3} & \dfrac{6EI}{L^2} & 0 & 0 \\[2mm] \dfrac{6EI}{L^2} & \dfrac{4EI}{L} & -\dfrac{6EI}{L^2} & \dfrac{2EI}{L} & 0 & 0 \\[2mm] -\dfrac{12EI}{L^3} & -\dfrac{6EI}{L^2} & \dfrac{36EI}{L} & \dfrac{6EI}{L^2} & -\dfrac{24EI}{L^3} & \dfrac{12EI}{L^2} \\[2mm] \dfrac{6EI}{L^2} & \dfrac{2EI}{L} & \dfrac{6EI}{L^2} & \dfrac{12EI}{L} & -\dfrac{12EI}{L^2} & \dfrac{4EI}{L} \\[2mm] 0 & 0 & -\dfrac{24EI}{L^3} & -\dfrac{12EI}{L^2} & \dfrac{24EI}{L^3} & -\dfrac{12EI}{L^2} \\[2mm] 0 & 0 & \dfrac{12EI}{L^2} & \dfrac{4EI}{L} & -\dfrac{12EI}{L^2} & \dfrac{8EI}{L} \end{bmatrix}^{-1}\times\begin{Bmatrix} -\dfrac{P}{2} \\[2mm] -\dfrac{PL}{8} \\[2mm] -\left(\dfrac{P}{2}+\dfrac{wL}{2}\right) \\[2mm] \left(\dfrac{PL}{8}-\dfrac{wL^2}{12}\right) \\[2mm] -\dfrac{wL}{2} \\[2mm] \dfrac{wL^2}{12} \end{Bmatrix}$$

The above relation may be condensed into following:

$$\begin{Bmatrix} \theta_2 \\ \theta_3 \end{Bmatrix} = \begin{bmatrix} \dfrac{12\,EI}{L} & \dfrac{4\,EI}{L} \\ \dfrac{4\,EI}{L} & \dfrac{8\,EI}{L} \end{bmatrix} \times \begin{Bmatrix} \dfrac{PL}{8} - \dfrac{wL^2}{12} \\ \dfrac{wL^2}{12} \end{Bmatrix} = \dfrac{L}{20\,EI}\begin{bmatrix} 2 & -1 \\ -1 & 3 \end{bmatrix}\begin{Bmatrix} \dfrac{PL}{8} - \dfrac{wL^2}{12} \\ \dfrac{wL^2}{12} \end{Bmatrix}$$

$$\begin{Bmatrix} \theta_2 \\ \theta_3 \end{Bmatrix} = \dfrac{L}{20\,EI}\begin{bmatrix} \dfrac{PL}{4} - \dfrac{wL^2}{4} \\ -\dfrac{PL}{8} + \dfrac{wL^2}{3} \end{bmatrix}$$

$$\theta_2 = \dfrac{PL^2}{80\,EI} - \dfrac{wL^3}{80\,EI}$$

$$\theta_3 = -\dfrac{PL}{160\,EI} - \dfrac{wL^3}{60\,EI}$$

Step 7: Determination of Member End Actions

The member end actions can be obtained from the corresponding member stiffness and the nodal displacements. The member end actions for each member are derived as shown below:

Member-(1):

$$\begin{Bmatrix} F_1 \\ M_1 \\ F_2 \\ M_2 \end{Bmatrix} = \dfrac{L}{20\,EI}\begin{bmatrix} \dfrac{12\,EI}{L^3} & \dfrac{6\,EI}{L^2} & -\dfrac{12\,EI}{L^3} & \dfrac{6\,EI}{L^2} \\ \dfrac{6\,EI}{L^2} & \dfrac{4\,EI}{L} & -\dfrac{6\,EI}{L^2} & \dfrac{2\,EI}{L} \\ -\dfrac{12\,EI}{L^3} & -\dfrac{6\,EI}{L^2} & \dfrac{12\,EI}{L^3} & -\dfrac{6\,EI}{L^2} \\ \dfrac{6\,EI}{L^2} & \dfrac{2\,EI}{L} & -\dfrac{6\,EI}{L^2} & \dfrac{4\,EI}{L} \end{bmatrix} \times \begin{Bmatrix} 0 \\ 0 \\ 0 \\ \dfrac{PL}{4} - \dfrac{wL^2}{4} \end{Bmatrix} = \begin{Bmatrix} \dfrac{3P}{40} - \dfrac{3wL}{40} \\ \dfrac{PL}{40} - \dfrac{wL^2}{40} \\ -\dfrac{3P}{40} + \dfrac{3wL}{40} \\ \dfrac{PL}{20} - \dfrac{wL^2}{20} \end{Bmatrix}$$

Member (2):

$$\begin{Bmatrix} F_2 \\ M_2 \\ F_3 \\ M_3 \end{Bmatrix} = \dfrac{EI}{L}\begin{bmatrix} \dfrac{24}{L^2} & \dfrac{12}{L} & -\dfrac{24}{L^2} & \dfrac{12}{L} \\ \dfrac{12}{L} & 8 & -\dfrac{12}{L} & 4 \\ -\dfrac{24}{L^2} & -\dfrac{12}{L} & \dfrac{24}{L^3} & -\dfrac{12}{L} \\ \dfrac{12}{L} & 4 & -\dfrac{12}{L} & 8 \end{bmatrix} \times \begin{Bmatrix} 0 \\ \dfrac{PL}{4} - \dfrac{wL^2}{4} \\ 0 \\ -\dfrac{PL}{8} + \dfrac{wL^2}{3} \end{Bmatrix} = \begin{Bmatrix} \dfrac{wL}{20} - \dfrac{3P}{40} \\ \dfrac{3PL}{40} - \dfrac{wL^2}{30} \\ -\dfrac{3wL}{20} - \dfrac{3P}{40} \\ \dfrac{wL^2}{12} \end{Bmatrix}$$

Actual member end actions:

Member (1):

$$
\begin{Bmatrix} \overline{F_1} \\ \overline{M_1} \\ \overline{F_2} \\ \overline{M_2} \end{Bmatrix} = \begin{Bmatrix} \dfrac{3P}{40} - \dfrac{3wL}{40} \\[2mm] \dfrac{PL}{40} - \dfrac{wL^2}{40} \\[2mm] -\dfrac{3P}{40} + \dfrac{3wL}{40} \\[2mm] \dfrac{PL}{20} - \dfrac{wL^2}{20} \end{Bmatrix} + \begin{Bmatrix} \dfrac{P}{2} \\[2mm] \dfrac{PL}{8} \\[2mm] \dfrac{P}{2} \\[2mm] -\dfrac{PL}{8} \end{Bmatrix} = \begin{Bmatrix} \dfrac{23P}{40} - \dfrac{3wL}{40} \\[2mm] \dfrac{6PL}{40} - \dfrac{wL^2}{40} \\[2mm] \dfrac{17P}{40} + \dfrac{3wL}{40} \\[2mm] -\dfrac{3PL}{40} - \dfrac{wL^2}{20} \end{Bmatrix}
$$

Member (2):

$$
\begin{Bmatrix} \overline{F_2} \\ \overline{M_2} \\ \overline{F_3} \\ \overline{M_3} \end{Bmatrix} = \begin{Bmatrix} \dfrac{wL}{20} + \dfrac{3P}{40} \\[2mm] \dfrac{3PL}{40} - \dfrac{wL^2}{30} \\[2mm] -\dfrac{wL}{20} - \dfrac{3P}{40} \\[2mm] \dfrac{wL^2}{12} \end{Bmatrix} + \begin{Bmatrix} \dfrac{wL}{2} \\[2mm] \dfrac{wL^2}{2} \\[2mm] \dfrac{wL}{2} \\[2mm] -\dfrac{wL^2}{12} \end{Bmatrix} = \begin{Bmatrix} \dfrac{11wL}{20} + \dfrac{3P}{40} \\[2mm] \dfrac{3PL}{40} + \dfrac{wL^2}{20} \\[2mm] \dfrac{9wL}{20} - \dfrac{3P}{40} \\[2mm] 0 \end{Bmatrix}
$$

The support reactions at the supports A, B and C are:

$$
\{F_x\} = \begin{Bmatrix} R_A \\ R_B \\ R_C \end{Bmatrix} = \begin{Bmatrix} \dfrac{23P}{40} - \dfrac{3wL}{40} \\[2mm] \dfrac{25wL}{40} + \dfrac{P}{2} \\[2mm] \dfrac{9wL}{20} - \dfrac{3P}{40} \end{Bmatrix}
$$

Putting the numerical values of L, P and w (P=15, L=4, w=4) the member actions and support reactions will be as follows:

Member End Actions:

$$
\begin{Bmatrix} F_2 \\ M_2 \\ F_3 \\ M_3 \end{Bmatrix} = \begin{Bmatrix} 9.925 \\ 7.7 \\ 6.075 \\ 0 \end{Bmatrix} \quad , \quad \begin{Bmatrix} F_1 \\ M_1 \\ F_2 \\ M_2 \end{Bmatrix} = \begin{Bmatrix} 7.425 \\ 7.4 \\ 7.575 \\ -7.7 \end{Bmatrix}
$$

Support reactions:

$$\{F_R\} = \begin{Bmatrix} R_A \\ R_B \\ R_C \end{Bmatrix} = \begin{Bmatrix} 7.425 \\ 17.5 \\ 6.075 \end{Bmatrix}$$

4.5 Plane Frame Analysis

The plane frame is a combination of plane truss and two dimensional beam. All the members lie in the same plane and are interconnected by rigid joints in case of plane frame.

The internal stress resultants at a cross-section of a plane frame member consist of axial force, bending moment and shear force.

Member Stiffness Matrix

In case of plane frame, the degrees of freedom at each node will be (i) axial deformation, (ii) vertical deformation and (iii) rotation. Thus the frame members have three degrees of freedom at each node as shown in the figure (a) below:

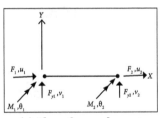

(a) Plane frame element

Therefore, the stiffness matrix of the frame in its local coordinate system will be the combination of 2-d truss and 2-d beam matrices:

$$[\bar{k}] = \begin{matrix} u_1 & v_1 & \theta_1 & u_2 & v_2 & \theta_2 \end{matrix}$$

$$[\bar{k}] = \begin{bmatrix} \dfrac{AE}{L} & 0 & 0 & -\dfrac{AE}{L} & 0 & 0 \\[2mm] 0 & \dfrac{12\,EI}{L^3} & \dfrac{6\,EI}{L^2} & 0 & -\dfrac{12\,EI}{L^3} & \dfrac{6\,EI}{L^2} \\[2mm] 0 & \dfrac{6\,EI}{L^2} & \dfrac{4\,EI}{L} & 0 & -\dfrac{6\,EI}{L^2} & \dfrac{2\,EI}{L} \\[2mm] -\dfrac{AE}{L} & 0 & 0 & \dfrac{AE}{L} & 0 & 0 \\[2mm] 0 & -\dfrac{12\,EI}{L^3} & -\dfrac{6\,EI}{L^2} & 0 & \dfrac{12\,EI}{L^3} & -\dfrac{6\,EI}{L^2} \\[2mm] 0 & \dfrac{6\,EI}{L^2} & \dfrac{2\,EI}{L} & 0 & -\dfrac{6\,EI}{L^2} & \dfrac{4\,EI}{L} \end{bmatrix} \qquad \ldots(1)$$

Generalized Stiffness Matrix

In plane frame the members are oriented in different directions and hence it is necessary to transform stiffness matrix of individual members from local to global co-ordinate system before formulating the global stiffness matrix by assembly.

The generalized stiffness matrix of a frame member can be obtained by transferring the matrix of local coordinate system into its global coordinate system. The transformation matrix can be expressed as:

$$[T] = \begin{bmatrix} \cos\theta & \sin\theta & 0 & 0 & 0 & 0 \\ -\sin\theta & \cos\theta & 0 & 0 & 0 & 0 \\ 0 & 0 & 1 & 0 & 0 & 0 \\ 0 & 0 & 0 & \cos\theta & \sin\theta & 0 \\ 0 & 0 & 0 & -\sin\theta & \cos\theta & 0 \\ 0 & 0 & 0 & 0 & 0 & 1 \end{bmatrix} \qquad ...(2)$$

Now, the generalized stiffness matrix of the member can be obtained from the relation of $[K] = [T]T[K][T]$. Thus considering $\lambda = \cos\theta$ and $\mu = \sin\theta$ the stiffness matrix in global coordinate system can be written as follows:

$$[K] = EI \begin{bmatrix} \lambda & -\mu & 0 & 0 & 0 & 0 \\ \mu & \lambda & 0 & 0 & 0 & 0 \\ 0 & 0 & 1 & 0 & 0 & 0 \\ 0 & 0 & 0 & \lambda & -\mu & 0 \\ 0 & 0 & 0 & \mu & \lambda & 0 \\ 0 & 0 & 0 & 0 & 0 & 1 \end{bmatrix} \times \begin{bmatrix} \dfrac{AE}{L} & 0 & 0 & -\dfrac{AE}{L} & 0 & 0 \\ 0 & \dfrac{12\,EI}{L^3} & \dfrac{6\,EI}{L^2} & 0 & -\dfrac{12\,EI}{L^3} & \dfrac{6\,EI}{L^2} \\ 0 & \dfrac{6\,EI}{L^2} & \dfrac{4\,EI}{L} & 0 & -\dfrac{6\,EI}{L^2} & \dfrac{2\,EI}{L} \\ -\dfrac{AE}{L} & 0 & 0 & \dfrac{AE}{L} & 0 & 0 \\ 0 & -\dfrac{12\,EI}{L^3} & -\dfrac{6\,EI}{L^2} & 0 & \dfrac{12\,EI}{L^3} & -\dfrac{6\,EI}{L^2} \\ 0 & \dfrac{6\,EI}{L^2} & \dfrac{2\,EI}{L} & 0 & -\dfrac{6\,EI}{L^2} & \dfrac{4\,EI}{L} \end{bmatrix}$$

$$\times \begin{bmatrix} \lambda & \mu & 0 & 0 & 0 & 0 \\ -\mu & \lambda & 0 & 0 & 0 & 0 \\ 0 & 0 & 1 & 0 & 0 & 0 \\ 0 & 0 & 0 & \lambda & \mu & 0 \\ 0 & 0 & 0 & -\mu & \lambda & 0 \\ 0 & 0 & 0 & 0 & 0 & 1 \end{bmatrix}$$

$$
=
\begin{bmatrix}
\left(\dfrac{EA}{L}\lambda^{2}+\dfrac{12EI}{L^{3}}\mu^{2}\right) & \left(\dfrac{EA}{L}\lambda\mu-\dfrac{12EI}{L^{3}}\lambda\mu\right) & -\dfrac{6EI}{L^{2}}\mu & \left(-\dfrac{EA}{L}\lambda^{2}-\dfrac{12EI}{L^{3}}\mu^{2}\right) & \left(-\dfrac{EA}{L}\lambda\mu+\dfrac{12EI}{L^{3}}\lambda\mu\right) & -\dfrac{6EI}{L^{2}}\mu \\[3mm]
\left(\dfrac{EA}{L}\lambda\mu-\dfrac{12EI}{L^{3}}\lambda\mu\right) & \left(\dfrac{EA}{L}\mu^{2}+\dfrac{12EI}{L^{3}}\lambda^{2}\right) & \dfrac{6EI}{L^{2}}\lambda & \left(-\dfrac{EA}{L}\lambda\mu+\dfrac{12EI}{L^{3}}\lambda\mu\right) & \left(-\dfrac{EA}{L}\mu^{2}-\dfrac{12EI}{L^{3}}\lambda\mu\right) & \dfrac{6EI}{L^{2}}\lambda \\[3mm]
-\dfrac{6EI}{L^{2}}\mu & \dfrac{6EI}{L^{2}}\lambda & \dfrac{4EI}{L} & \dfrac{6EI}{L^{2}}\mu & -\dfrac{6EI}{L^{2}}\lambda & \dfrac{2EI}{L} \\[3mm]
\left(-\dfrac{EA}{L}\lambda^{2}-\dfrac{12EI}{L^{3}}\mu^{2}\right) & \left(-\dfrac{EA}{L}\lambda\mu+\dfrac{12EI}{L^{3}}\lambda\mu\right) & \dfrac{6EI}{L^{2}}\mu & \left(\dfrac{EA}{L}\lambda^{2}+\dfrac{12EI}{L^{3}}\mu^{2}\right) & \left(\dfrac{EA}{L}\lambda\mu-\dfrac{12EI}{L^{3}}\lambda^{2}\right) & \dfrac{6EI}{L^{2}}\mu \\[3mm]
\left(-\dfrac{EA}{L}\lambda\mu+\dfrac{12EI}{L^{3}}\lambda\mu\right) & \left(-\dfrac{EA}{L}\mu^{2}+\dfrac{12EI}{L^{3}}\lambda^{2}\right) & -\dfrac{6EI}{L^{2}}\lambda & \left(\dfrac{EA}{L}\lambda\mu-\dfrac{12EI}{L^{3}}\lambda\mu\right) & \left(\dfrac{EA}{L}\mu^{2}+\dfrac{12EI}{L^{3}}\lambda^{2}\right) & -\dfrac{6EI}{L^{2}}\lambda \\[3mm]
-\dfrac{6EI}{L^{2}}\mu & \dfrac{6EI}{L^{2}}\lambda & \dfrac{2EI}{L} & \dfrac{6EI}{L^{2}}\mu & -\dfrac{6EI}{L^{2}}\lambda & \dfrac{4EI}{L}
\end{bmatrix}
$$

Problem:

1. Let us analyze the plane frame shown below. Assume the modulus of elasticity of the horizontal member is 1.5 times that of the vertical member and length of the vertical member is 1.5 times that of horizontal member. And also let us determine the bending moment and reactions at support assuming the length, cross section area and modulus of elasticity of vertical member as 3.0 m, 0.4 x 0.4 m2 and 2 x 1011N/mm2, respectively.

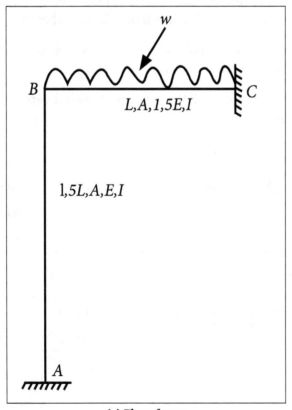

(a) Plane frame

Solution

Given:

Step 1: Numbering of Nodes and Members

The numbering of members and joints of the plane frame are as shown below:

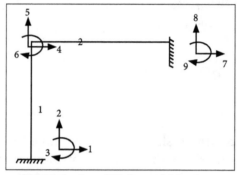

(b) Numbering of Nodes and Members.

The members AB and BC are designated as (1) and (2). The points A, B and C are designated by nodes 1, 2 and 3. The member information for the frame is shown in tabulated form as shown in the table (a). The coordinate of node 1 is assumed as (0, 0). The coordinate and restraint joint information are shown in the table (b).

The integer 1 in the restraint list indicates the restraint exists and 0 indicates the restraint at that particular direction does not exist.

Thus, in node no. 2, the integer 0 all the restraint type indicates that the joint is in free all the three directions.

(a) Member Information for Beam

Member number	Starting node	Ending node	Rigidity modulus
1	1	2	EI
2	2	3	1.5EI

(b) Nodal Information for Beam

Node No.	Coordinates		Restraint List		
	x	y	Axial	Vertical	Rotation
1	0	0	1	1	1
2	0	1.5L	0	0	0
3	L	1.5L	1	1	1

Step 2: Formation of Member Stiffness Matrix

The individual member stiffness matrices can be found out directly from equation shown above.

Thus the stiffness matrices of each member in global coordinate system are given below based on their individual member properties and orientations. Thus the stiffness matrix of member (1) is given by:

$$[k]_1 = \begin{array}{c c c c c c c}
& 1 & 2 & 3 & 4 & 5 & 6 & \\
\begin{bmatrix} \dfrac{12\,EI}{(1.5L)^3} & 0 & -\dfrac{6\,EI}{(1.5L)^2} & -\dfrac{12\,EI}{(1.5L)^3} & 0 & -\dfrac{6\,EI}{(1.5L)^2} \\[2ex]
0 & \dfrac{AE}{(1.5L)} & 0 & 0 & -\dfrac{AE}{(1.5L)} & 0 \\[2ex]
-\dfrac{6\,EI}{(1.5L)^2} & 0 & \dfrac{4\,EI}{(1.5L)} & \dfrac{6\,EI}{(1.5L)^2} & 0 & \dfrac{2\,EI}{1.5L} \\[2ex]
-\dfrac{12\,EI}{(1.5L)^2} & 0 & \dfrac{6\,EI}{(1.5L)^2} & \dfrac{12\,EI}{(1.5L)^3} & 0 & \dfrac{6\,EI}{(1.5L)^2} \\[2ex]
0 & -\dfrac{AE}{(1.5L)} & 0 & 0 & \dfrac{AE}{(1.5L)} & 0 \\[2ex]
-\dfrac{6\,EI}{(1.5L)^2} & 0 & \dfrac{2\,EI}{(1.5L)} & \dfrac{6\,EI}{(1.5L)^2} & 0 & \dfrac{4\,EI}{(1.5L)} \end{bmatrix} & \begin{matrix} 1 \\[2ex] 2 \\[2ex] 3 \\[2ex] 4 \\[2ex] 5 \\[2ex] 6 \end{matrix}
\end{array}$$

Similarly, the stiffness matrix of member (2) is:

$$[k]_2 = \begin{array}{c c c c c c c}
& 4 & 5 & 6 & 7 & 8 & 9 & \\
\begin{bmatrix} \dfrac{A(1.5E)}{L} & 0 & 0 & -\dfrac{A(1.5E)}{L} & 0 & 0 \\[2ex]
0 & \dfrac{12(1.5E)I}{L^3} & \dfrac{6(1.5E)I}{L^2} & 0 & \dfrac{12(1.5E)I}{L^3} & \dfrac{6(1.5E)I}{L^2} \\[2ex]
0 & \dfrac{6(1.5E)I}{L^2} & \dfrac{4(1.5E)I}{L} & 0 & -\dfrac{6(1.5E)I}{L^2} & \dfrac{2(1.5E)I}{L} \\[2ex]
-\dfrac{A(1.5E)}{L} & 0 & 0 & \dfrac{A(1.5E)}{L} & 0 & 0 \\[2ex]
0 & -\dfrac{12(1.5E)I}{L^3} & -\dfrac{6(1.5E)I}{L^2} & 0 & \dfrac{12(1.5E)I}{L^3} & -\dfrac{6(1.5E)I}{L^2} \\[2ex]
0 & \dfrac{6(1.5E)I}{L^2} & \dfrac{2(1.5E)I}{L} & 0 & -\dfrac{6(1.5E)I}{L^2} & \dfrac{4(1.5E)I}{L} \end{bmatrix} & \begin{matrix} 4 \\[2ex] 5 \\[2ex] 6 \\[2ex] 7 \\[2ex] 8 \\[2ex] 9 \end{matrix}
\end{array}$$

Step 3: Formulation of Global Stiffness Matrix

The global stiffness matrix is obtained by assembling by assembling the local stiffness matrix of member (1) and (2) as follows:

$$
[K] = \begin{bmatrix}
\dfrac{32\,EI}{9L^2} & 0 & -\dfrac{8\,EI}{3L^2} & -\dfrac{32\,EI}{9L^2} & 0 & -\dfrac{8\,EI}{3L^2} & 0 & 0 & 0 \\[6pt]
0 & \dfrac{2AE}{3L} & 0 & 0 & -\dfrac{2AE}{3L} & 0 & 0 & 0 & 0 \\[6pt]
\dfrac{8\,EI}{3L^2} & 0 & \dfrac{8\,EI}{3L} & \dfrac{8\,EI}{3L^2} & 0 & \dfrac{4\,EI}{3L} & 0 & 0 & 0 \\[6pt]
-\dfrac{32\,EI}{9L^2} & 0 & \dfrac{8\,EI}{3L^2} & \left(\dfrac{32\,EI}{9L^2}+\dfrac{1.5\,EA}{L}\right) & 0 & \dfrac{8\,EI}{3L^2} & -\dfrac{1.5\,EA}{L} & 0 & 0 \\[6pt]
0 & -\dfrac{2AE}{3L} & 0 & 0 & \left(\dfrac{2AE}{3L}+\dfrac{18\,EI}{L^2}\right) & \dfrac{9\,EI}{L^2} & 0 & -\dfrac{18\,EI}{L^2} & \dfrac{9\,EI}{L^2} \\[6pt]
-\dfrac{8\,EI}{3L^2} & 0 & \dfrac{4\,EI}{3L^2} & \dfrac{8\,EI}{3L^2} & \dfrac{9\,EI}{L^2} & \left(\dfrac{8\,EI}{3L^2}+\dfrac{6\,EI}{L}\right) & 0 & \dfrac{9\,EI}{L^2} & \dfrac{3\,EI}{L} \\[6pt]
0 & 0 & 0 & -\dfrac{1.5\,AE}{L} & 0 & 0 & \dfrac{1.5\,EA}{L} & 0 & 0 \\[6pt]
0 & 0 & 0 & 0 & -\dfrac{18\,EI}{L^2} & -\dfrac{9\,EI}{L^2} & 0 & \dfrac{18\,EI}{L^2} & -\dfrac{9\,EI}{L^2} \\[6pt]
0 & 0 & 0 & 0 & \dfrac{9\,EI}{L^2} & \dfrac{3\,EI}{L} & 0 & -\dfrac{9\,EI}{L^2} & \dfrac{6\,EI}{L}
\end{bmatrix}
\begin{matrix}1\\2\\3\\4\\5\\6\\7\\8\\9\end{matrix}
$$

Step 4: Boundary Conditions

The boundary conditions according to the support of the frame can be expressed in terms of the displacement vector. The displacement vector will be as follows:

$$
\{d\} = \begin{bmatrix}
0 \\
0 \\
0 \\
\delta x_B \\
\delta y_B \\
\theta_B \\
0 \\
0 \\
0
\end{bmatrix}
$$

Here, δx_B, δy_B and θ_B indicate the displacement in X-direction, displacement in Y-direction and rotation at point B.

Step 5: Load Vector

The distributed load on member (2) can be replaced by its equivalent joint load as shown in the figure below:

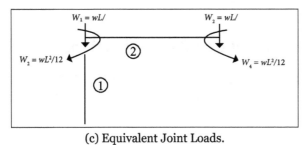

(c) Equivalent Joint Loads.

Thus, the equivalent joint load vector can be written as:

$$\{F\}=\begin{bmatrix} 0 \\ 0 \\ 0 \\ 0 \\ -\dfrac{wL}{2} \\ -\dfrac{wL^2}{12} \\ 0 \\ -\dfrac{wL}{2} \\ \dfrac{wL^2}{12} \end{bmatrix}$$

Step 6: Determination of Unknown Displacements

The unknown displacements can be obtained from the relationship of $\{F\} = [K]\{d\}$ or $\{d\} =[k]$-1$\{F\}$. Now eliminating the rows and columns in the stiffness matrix and force matrix, corresponding to zero elements in displacement matrix, the reduced matrix will be as follows:

$$\begin{bmatrix} \delta x_B \\ \delta y_B \\ \theta_B \end{bmatrix} = \begin{bmatrix} \left(\dfrac{32EI}{9L^3}+\dfrac{1.5EA}{L}\right) & 0 & \dfrac{8EI}{3L^2} \\ 0 & \left(\dfrac{2AE}{3L}+\dfrac{18EI}{L^3}\right) & \dfrac{9EI}{L^2} \\ \dfrac{8EI}{3L^2} & \dfrac{9EI}{3L^2} & \left(\dfrac{8EI}{3L}+\dfrac{6EI}{L}\right) \end{bmatrix}^{-1} \begin{bmatrix} 0 \\ -\dfrac{wL}{2} \\ -\dfrac{wL^2}{12} \end{bmatrix}$$

Thus, the unknown displacements will be:

$$\begin{bmatrix} \delta x_B \\ \delta y_B \\ \theta_B \end{bmatrix} = \frac{1}{10^{10}} \begin{bmatrix} 0.04327\,w \\ -1.7127\,w \\ -5.4978\,w \end{bmatrix}$$

Step 7: Determination of member end actions

The member end actions can be obtained from the corresponding member stiffness and the nodal displacements. The member end actions for each member are derived as shown below:

Member – (1)

In case of member (1), the member forces will be: $\{F_m\}_1 = [K]_{(1)}\{d\}_{(1)}$

$$\begin{bmatrix} F_{x1} \\ F_{y1} \\ M_1 \\ F_{x2} \\ F_{y2} \\ M_2 \end{bmatrix} = 10^6 \begin{bmatrix} 56.17 & 0 & -126.4 & -56.17 & 0 & -126.4 \\ 0 & 7110 & 0 & 0 & -7110 & 0 \\ -126.4 & 0 & 379.2 & 126.4 & 0 & 189.6 \\ -56.17 & 0 & 379.2 & 56.17 & 0 & 126.4 \\ 0 & -7110 & 0 & 0 & 7110 & 0 \\ -126.4 & 0 & 189.6 & 126.4 & 0 & 379.2 \end{bmatrix}$$

$$\times \begin{bmatrix} 0 \\ 0 \\ 0 \\ 4.327\times10^{-12}\,w \\ -1.7127\times10^{-10}\,w \\ -5.4978\times10^{-10}\,w \end{bmatrix}$$

$$= \begin{bmatrix} 0.0697\,w \\ 1.2177\,w \\ -0.10479\,w \\ -0.06925\,w \\ -1.21661\,w \\ -0.20793w \end{bmatrix}$$

It is to be noted that {Fm} are the end actions due to joint loads. Hence it must be added to the corresponding end actions in the restrained structure in order to obtain the end actions due to the loads. Therefore, {Fm}actual are the true member end actions due to actual loading system can be expressed as:

$$\{F_m\}_{actual} = \{F_m\} + \{F_{fm}\}$$

Where, $\{F_{fm}\}$ are the end actions in the restrained structure. Since there is no load acting on member (1), the actual end actions will be:

$$\{F_m\}_{actual} = \begin{bmatrix} 0.0697\,w \\ 1.2177\,w \\ -0.10479\,w \\ -0.06925\,w \\ -1.21661\,w \\ -0.20793w \end{bmatrix} + \begin{bmatrix} 0 \\ 0 \\ 0 \\ 0 \\ 0 \\ 0 \end{bmatrix} = \begin{bmatrix} 0.0697\,w \\ 1.2177\,w \\ -0.10479\,w \\ -0.06925\,w \\ -1.21661\,w \\ -0.20793w \end{bmatrix}$$

Member (2)

In similar way, the member forces in member (2) will be $\{Fm\}_{(2)} = [K]_{(2)}\{d\}_{(2)}$

$$\begin{bmatrix} F_{x2} \\ F_{y2} \\ M_2 \\ F_{x3} \\ F_{y3} \\ M_3 \end{bmatrix} = 10^9 \begin{bmatrix} 16 & 0 & 0 & -16 & 0 & 0 \\ 0 & 0.284 & 0.426 & 0 & -0.284 & 0.426 \\ 0 & 0.426 & 0.853 & 0 & -0.426 & 0.426 \\ -16 & 0 & 0 & 16 & 0 & 0 \\ 0 & -0.284 & -0.426 & 0 & 0.284 & -0.426 \\ 0 & 0.426 & 0.426 & 0 & -0.426 & 0.853 \end{bmatrix} \times \begin{bmatrix} 4.327\times10^{-12}\,w \\ -1.7127\times10^{-10}\,w \\ -5.4978\times10^{-10}\,w \\ 0 \\ 0 \\ 0 \end{bmatrix}$$

$$= \begin{bmatrix} 0.069232\,w \\ -0.28325\,w \\ -0.54215\,w \\ -0.06923\,w \\ 0.283245\,w \\ -0.3076\,w \end{bmatrix}$$

The actual member forces in the member (2) will be:

$$\{F_m\}_{actual} = \begin{bmatrix} 0.069232\,w \\ -0.28325\,w \\ -0.54215\,w \\ -0.06923\,w \\ 0.283245\,w \\ -0.3076\,w \end{bmatrix} + \begin{bmatrix} 0 \\ 1.5w \\ 0.75w \\ 0 \\ 1.5w \\ -0.75w \end{bmatrix} = \begin{bmatrix} 0.0692\,w \\ 1.2167\,w \\ 0.2078\,w \\ -0.0692\,w \\ 1.7832\,w \\ -1.0576\,w \end{bmatrix}$$

4.6 Analysis of Grid and Space Frame

Analysis of Grid

A grid is a structure on which loads are applied perpendicular to the plane of the structure, as opposed to a plane frame, where loads are applied in the plane of the structure.

We will now develop the grid element stiffness matrix. The elements of a grid are assumed to be rigidly connected, so that the original angles between elements connected together at a node remain unchanged. Both torsional and bending moment continuity then exist at the node point of a grid.

Examples of grids include floor and bridge deck systems:

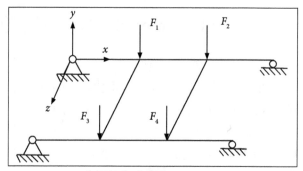

(a) Typical grid structure.

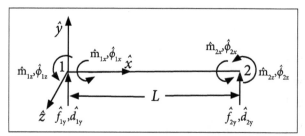

(b) Grid element with nodal degrees of freedom and nodal forces.

A typical grid structure subjected to loads F_1, F_2, F_3 and F_4 is shown in the figure (a). We will now consider the development of the grid element stiffness matrix and element equations.

A representative grid element with the nodal degrees of freedom and nodal forces is shown in the figure (b). The degrees of freedom at each node for a grid are a vertical deflection $\hat{d}_{iy}$ (normal to the grid), a torsional rotation $\hat{\phi}_{ix}$ about the $\hat{x}$ axis and a bending rotation $\hat{\phi}_{iz}$ about the $\hat{z}$ axis.

Any effect of axial displacement is ignored; that is, $\hat{d}_{ix} = 0$. The nodal forces consist of a transverse force $\hat{f}_{iy}$, a torsional moment $\hat{m}_{ix}$ about the axis and a bending moment $\hat{m}_{iz}$ about the $\hat{z}$ axis. Grid elements do not resist axial loading, that is $\hat{f}_{ix} = 0$.

$$\hat{\underline{k}} = \frac{EI}{L^3} \begin{bmatrix} 12 & 6L & -12 & 6L \\ 6L & 4L^2 & -6L & 2L^2 \\ -12 & -6L & 12 & -6L \\ 6L & 2L^2 & -6L & 4L^2 \end{bmatrix} \qquad ...(1)$$

To develop the local stiffness matrix for a grid element, we need to include the torsional effects in the basic beam element stiffness matrix equation.

We can derive the torsional bar element stiffness matrix in a manner analogous to that used for the axial bar element stiffness matrix. In the derivation, we simply replace $\hat{f}_{ix}$ with $\hat{m}_{ix}$, $\hat{d}_{ix}$ with $\hat{\phi}_{ix}$, E with G(the shear modulus),A with J(the torsional constant or stiffness factor), σ with τ(shear stress) and ε with γ(shear strain).

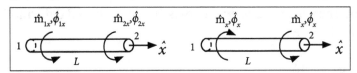

(c) Nodal and element torque sign conventions.

The actual derivation is briefly presented as follows. We assume a circular cross section with radius R for simplicity but without loss of generalization.

Step 1: Figure (c) shows the sign conventions for nodal torque and angle of twist and for element torque.

Step 2: We assume a linear angle-of-twist variation along the $\hat{x}$ axis of the bar such that:

$$\hat{\phi} = a_1 = a_2 \hat{x} \qquad ...(2)$$

Using the usual procedure of expressing a1 and a2 in terms of unknown nodal angles of twist $\hat{\phi}_{1x}$ and $\hat{\phi}_{2x}$, we obtain:

$$\hat{\phi} = \left(\frac{\hat{\phi}_{2x} - \hat{\phi}_{1x}}{L} \right) \hat{x} + \hat{\phi}_{1x} \qquad ...(3)$$

or, in matrix form, equation(3) becomes:

$$\hat{\phi} = \begin{bmatrix} N_1 & N_2 \end{bmatrix} \begin{Bmatrix} \hat{\phi}_{1x} \\ \hat{\phi}_{2x} \end{Bmatrix} \qquad ...(4)$$

With the shape functions given by:

$$N_1 = 1 - \frac{\hat{x}}{L} \qquad N_2 = \frac{\hat{x}}{L} \qquad ...(5)$$

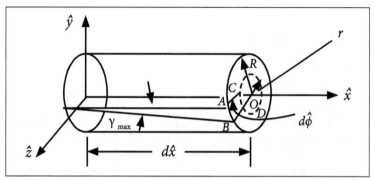

(d) Torsional deformation of a bar segment.

Step 3: We obtain the shear strain γ angle of twist $\hat{\phi}$ relationship by considering the torsional deformation of the bar segment shown in the figure (d). Assuming that all radial lines, such as OA, remain straight during twisting or torsional deformation, we observe that are length is given by:

$$\widehat{AB} = \gamma_{max}\, d\hat{x} = R\, d\hat{\phi}$$

Solving for the maximum shear strain γmax, we obtain:

$$\gamma_{max} = \frac{R\, d\hat{\phi}}{d\hat{x}}$$

Similarly, at any radial position r, we then have, from similar triangles OAB and OC:

$$\gamma = r\frac{d\hat{\phi}}{d\hat{x}} = \frac{r}{L}\left(\hat{\phi}_{2x} - \hat{\phi}_{1x}\right) \qquad\qquad ...(6)$$

Where we have used equation (3) to derive the final expression in equation (6).

The shear stress τ/shear strain γ relationship for linear-elastic isotropic materials is given by:

$$\tau = G\gamma \qquad\qquad ...(7)$$

Where, G-Shear modulus of the material.

Step 4: We derive the element stiffness matrix in the following manner. From elementary mechanics, we have the shear stress related to the applied torque by:

$$\hat{m}_x = \frac{\tau J}{R} \qquad\qquad ...(8)$$

Where, J is called the polar moment of inertia for the circular cross section or generally, the torsional constant for noncircular cross sections. Using equations (6) and (7) in equation (8), we obtain:

$$\hat{m}_x = \frac{GJ}{L}\left(\hat{\phi}_{2x} - \hat{\phi}_{1x}\right) \qquad ...(9)$$

By the nodal torque sign convention of the figure (d):

$$\hat{m}_{1x} = -\hat{m}_x \qquad ...(10)$$

Or by using equation (9) in equation (10), we obtain:

$$\hat{m}_{1x} = \frac{GJ}{L}\left(\hat{\phi}_{1x} - \hat{\phi}_{2x}\right) \qquad ...(11)$$

Similarly,

$$\hat{m}_{2x} = \hat{m}_x \qquad ...(12)$$

Or

$$\hat{m}_{2x} = \frac{GJ}{L}\left(\hat{\phi}_{2x} - \hat{\phi}_{1x}\right) \qquad ...(13)$$

Expressing equations (11) and (13) together in matrix form, we have the resulting torsion bar stiffness matrix equation:

$$\begin{Bmatrix} \hat{m}_{1x} \\ \hat{m}_{2x} \end{Bmatrix} = \frac{GJ}{L}\begin{bmatrix} 1 & -1 \\ -1 & 1 \end{bmatrix}\begin{Bmatrix} \hat{\phi}_{1x} \\ \hat{\phi}_{2x} \end{Bmatrix} \qquad ...(14)$$

Hence, the stiffness matrix for the torsion bar is given by:

$$\hat{\underline{k}} = \frac{GJ}{L}\begin{bmatrix} 1 & -1 \\ -1 & 1 \end{bmatrix} \qquad ...(15)$$

The cross sections of various structures, such as bridge decks, are often not circular. However, equations (14) and (15) are still general, to apply them to other cross sections, we simply evaluate the torsional constant J for the particular cross section.

For instance, for cross sections made up of thin rectangular shapes such as channels, angles or 1 shape, we approximate J by:

$$J = \sum \frac{1}{3}b_i t^3{}_i \qquad ...(16)$$

Where, bi is the length of any element of the cross section and ti is the thickness of any element of the cross section.

(a) Torsional constants J and shear centers SC for various cross sections:

Cross Section	Torsional Constant
1. Channel 	$J = \dfrac{t^3}{3}(h + 2b)$ $e = \dfrac{h^2 b^2 t}{4I}$
2. Angle 	$J = \dfrac{1}{3}\left(b_1\, t_1^3 + b_2\, t_2^3\right)$
3. Z section 	$J = \dfrac{t^3}{3}(2b + h)$
4. Wide-flanged beam with unequal flanges 	$J = \dfrac{1}{3}\left(b_1 t_1^3 + b_2\, t_2^3 + h\, t_w^3\right)$
5. Solid circular 	$J = \dfrac{\pi}{2} r^4$
6. Closed hollow rectangular 	$J = \dfrac{2 t t_1 (a - t)^2 (b - t_1)^2}{a\,t + b\,t_1 - t^2 - t_1^2}$

In the table (a), we list values of J for various common cross sections. The first four cross sections are called open sections. Equation (16) applies only to these open cross sections.

We assume the loading to go through the shear center of these open cross sections in order to prevent twisting of the cross section.

$$
\begin{Bmatrix} \hat{f}_{1y} \\ \hat{m}_1 \\ \hat{f}_{2y} \\ \hat{m}_2 \end{Bmatrix} = \frac{EI}{L^3} \begin{bmatrix} 12 & 6L & -12 & 6L \\ 6L & 4L^2 & -6L & 2L^2 \\ -12 & -6L & 12 & -6L \\ 6L & 2L^2 & -6L & 4L^2 \end{bmatrix} \begin{Bmatrix} \hat{d}_{1y} \\ \hat{\phi}_1 \\ \hat{d}_{2y} \\ \hat{\phi}_2 \end{Bmatrix} \quad ...(a)
$$

On combining the torsional effects of equation(4) with the shear and bending effects of equation (a), we obtain the local stiffness matrix equation for a grid element as:

$$
\begin{Bmatrix} \hat{f}_{1y} \\ \hat{m}_{1x} \\ \hat{m}_{1z} \\ \hat{f}_{2y} \\ \hat{m}_{2x} \\ \hat{m}_{2z} \end{Bmatrix} = \begin{bmatrix} \dfrac{12\,EI}{L^3} & 0 & \dfrac{6\,EI}{L^2} & \dfrac{-12\,EI}{L^3} & 0 & \dfrac{6\,EI}{L^2} \\ & \dfrac{GJ}{L} & 0 & 0 & \dfrac{-GJ}{L} & 0 \\ & & \dfrac{4\,EI}{L} & \dfrac{-6\,EI}{L^2} & 0 & \dfrac{2\,EI}{L} \\ & & & \dfrac{12\,EI}{L^3} & 0 & \dfrac{-6\,EI}{L^2} \\ & & & & \dfrac{GJ}{L} & 0 \\ \text{Symmetry} & & & & & \dfrac{4\,EI}{L} \end{bmatrix} \begin{Bmatrix} \hat{d}_{1y} \\ \hat{\phi}_{1x} \\ \hat{\phi}_{1z} \\ \hat{d}_{2y} \\ \hat{\phi}_{2x} \\ \hat{\phi}_{2z} \end{Bmatrix} \quad ...(17)
$$

Where, from equation (17), the local stiffness matrix for a grid element is:

$$
\hat{\underline{k}}_G = \begin{bmatrix} \dfrac{12\,EI}{L^3} & 0 & \dfrac{6\,EI}{L^2} & \dfrac{-12\,EI}{L^3} & 0 & \dfrac{6\,EI}{L^2} \\ 0 & \dfrac{GJ}{L} & 0 & 0 & \dfrac{-GJ}{L} & 0 \\ \dfrac{6\,EI}{L^2} & 0 & \dfrac{4\,EI}{L} & \dfrac{-6\,EI}{L^2} & 0 & \dfrac{2\,EI}{L} \\ \dfrac{-12\,EI}{L^3} & 0 & \dfrac{-6\,EI}{L^2} & \dfrac{12\,EI}{L^3} & 0 & \dfrac{-6\,EI}{L^2} \\ 0 & \dfrac{-GJ}{L} & 0 & 0 & \dfrac{GJ}{L} & 0 \\ \dfrac{6\,EI}{L^2} & 0 & \dfrac{2\,EI}{L} & \dfrac{-6\,EI}{L^2} & 0 & \dfrac{4\,EI}{L} \end{bmatrix} \quad ...(18)
$$

And the degrees of freedom are in the order (1) vertical deflection, (2) torsional rotation and (3) bending rotation, as indicated by the notation used above the columns of equation (18).

The transformation matrix relating local to global degrees of freedom for a grid is given by:

$$\underline{T}_G = \begin{bmatrix} 1 & 0 & 0 & 0 & 0 & 0 \\ 0 & C & S & 0 & 0 & 0 \\ 0 & -S & C & 0 & 0 & 0 \\ 0 & 0 & 0 & 1 & 0 & 0 \\ 0 & 0 & 0 & 0 & 0 & S \\ 0 & 0 & 0 & 0 & -S & C \end{bmatrix} \quad \text{...(19)}$$

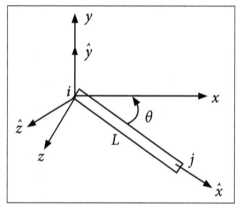

(e) Grid element arbitrarily oriented in the x-z plane.

Where, θ is now positive, taken counterclockwise from x to $\hat{x}$ in the x-z plane Figure (e).

and,

$$C = \cos\theta = \frac{x_j - x_i}{L} \qquad S = \sin\theta = \frac{z_j - z_i}{L}$$

Where, L is the length of the element from node i to node j. As indicated by equation (19) for a grid, the vertical deflection d_y is invariant with respect to a coordinate transformation (that is $y = \hat{y}$) (Figure (e)).

The global stiffness matrix for a grid element arbitrarily oriented in the x-z plane is then given by using equations (18) and (19) in,

The global stiffness matrix for a grid element arbitrarily oriented in the x-z plane is then given by using equations (18) and (19) in:

$$\underline{k}_G = \underline{T}_G^T \, \hat{\underline{k}}_G \, \underline{T}_G \quad \text{...(20)}$$

Now that we have formulated the global stiffness matrix for the grid element, the procedure for solution then follows in the same manner as that for the plane frame.

Space Frame

A space frame or space truss is a three dimensional assemblage of line members, each member being joined at its ends, either to the foundation or to other members by frictionless ball-and-socket joints.

The simplest space frame consists of six members joined to form a tetrahedron. Beginning with a six member tetrahedron, a stable space frame can be constructed by successive addition of three new members and a new joint.

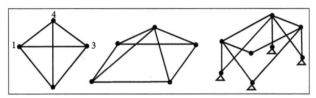

Basic tetrahedron Pentahedron Utility space frame.

In space trusses, the relationship between the number of members (m) and number of joints (j) given by:

$$m = (3j - 6)$$

m -Gives the minimum number of members required to build a stable trusses or frame.

m =3j (if the frame starts from a rigid base)

If a truss or frame has less number of members than what is given by the above equations the truss or frame will be unstable. If it has more members, the frame will be internally statically indeterminate.

The geometry of space structures makes their analysis slightly more complex compared to two-dimensional structures.

Space frames are an increasingly common architectural technique especially for large roof spans in commercial and industrial buildings. The rigid jointed frames such as building frames are usually three dimensional space structures.

Thus in case of certain structures like multistoried buildings, it is necessary consider 3-dimensional effects for analysis. The space frame constitutes the final step of increasing complexity.

It consists of plane frame and grid actions. The displacement and rotation vector associated with each joint have three components in case of space frame structures.

There are six equilibrium equations associated with each joint. The degrees of freedom at each node of the space frame member will be (i) displacement in three perpendicular directions and (ii) rotations in three different directions.

Therefore, the degrees of freedom in each node of the member will be six as shown in the figure below. The stiffness matrix in local coordinate system considering all possible degrees of freedom will be as given in the table below:

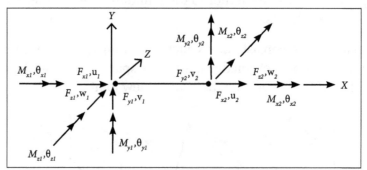

(1) Degrees of freedom for space frame member.

Stiffness Matrix of Space Frame Member

	1	2	3	4	5	6	7	8	9	10	11	12
1	$\dfrac{EA_x}{L}$	0	0	0	0	0	$-\dfrac{EA_x}{L}$	0	0	0	0	0
2	0	$\dfrac{12EI_z}{L^3}$	0	0	0	$\dfrac{6EI_z}{L^2}$	0	$-\dfrac{12EI_z}{L^3}$	0	0	0	$\dfrac{6EI_z}{L^2}$
3	0	0	$\dfrac{12EI_x}{L^3}$	0	$-\dfrac{6EI_y}{L^2}$	0	0	0	$-\dfrac{12EI_y}{L^3}$	0	$-\dfrac{6EI_y}{L^2}$	0
4	0	0	0	$\dfrac{GL_x}{L}$	0	0	0	0	0	$-\dfrac{GL_x}{L}$	0	0
5	0	0	$-\dfrac{6EI_x}{L^2}$	0	$\dfrac{4EI_y}{L}$	0	0	0	$\dfrac{6EI_y}{L^2}$	0	$\dfrac{2EI_y}{L}$	0
6	0	$\dfrac{6EI_z}{L^2}$	0	0	0	$\dfrac{4EI_z}{L}$	0	$-\dfrac{6EI_z}{L^2}$	0	0	0	$\dfrac{2EI_z}{L}$
7	$-\dfrac{EA_x}{L}$	0	0	0	0	0	$\dfrac{EA_x}{L}$	0	0	0	0	0
8	0	$-\dfrac{12EI_z}{L^3}$	0	0	0	$-\dfrac{6EI_z}{L^2}$	0	$\dfrac{12EI_z}{L^3}$	0	0	0	$-\dfrac{6EI_z}{L^2}$
9	0	0	$-\dfrac{12EI_y}{L^3}$	0	$\dfrac{6EI_y}{L^2}$	0	0	0	$\dfrac{12EI_y}{L^3}$	0	$\dfrac{6EI_y}{L^2}$	0
10	0	0	0	$-\dfrac{GL_x}{L}$	0	0	0	0	0	$\dfrac{GL_x}{L}$	0	0
11	0	0	$-\dfrac{6EI_y}{L^2}$	0	$\dfrac{2EI_y}{L}$	0	0	0	$\dfrac{6EI_y}{L^2}$	0	$\dfrac{4EI_y}{L}$	0
12	0	$\dfrac{6EI_z}{L^2}$	0	0	0	$\dfrac{2EI_z}{L}$	0	$-\dfrac{6EI_z}{L}$	0	0	0	$\dfrac{4EI_z}{L}$

The generalized stiffness matrix of a rigid jointed space frame member can be obtained by transferring the matrix of local coordinate system into its global coordinate system.

The transformation matrix will become a square matrix of size 12×12 in this case as the degrees of freedom for each node/joint is six.

4.7 Introduction to Plate Bending Problems

A plate can be considered the two-dimensional extension of a beam in simple bending. Both beams and plates support loads transverse or perpendicular to their plane and through bending action.

A plate is flat (if it were curved, it would become a shell). A beam has a single bending moment resistance, while a plate resists bending about two axes and has a twisting moment.

We will consider the classical thin-plate theory or Kirchhoff plate theory. Many of the assumptions of this theory are analogous to the classical beam theory or Euler–Bernoulli beam theory.

Basic Behavior of Geometry and Deformation

We begin the derivation of the basic thin-plate equations by considering the thin plate in the x-y plane and of thickness t measured in the z direction shown in the figure (a).

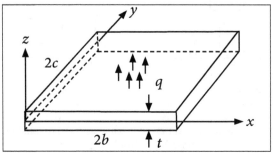

(a) Basic thin plate showing transverse loading and dimensions.

The plate surfaces are at $z = \pm t/2$ and its mid surface is at $z = 0$. The assumed basic geometry of the plate is as follows:

- The plate thickness is much smaller than its in-plane dimensions b and c(that is, t<<b or c). (If t is more than about one-tenth the span of the plate, then transverse shear deformation must be accounted for and the plate is then said to be thick.)

- The deflection w is much less than the thickness t (that is, w/t<<1).

Kirchhoff Assumptions

Consider a differential slice cut from the plate by planes perpendicular to the x axis as shown in the figure 1(a). Loading q causes the plate to deform laterally or upward in the z direction and the deflection w of point P is assumed to be a function of x and y only, that is, w = w(x, y) and the plate does not stretch in the z direction.

A line a-b drawn perpendicular to the plate surfaces before loading remains perpendicular to the surfaces after loading [Figure 1(b)].

This is consistent with the Kirchhoff assumptions as follows:

- Normal remains normal. This implies that transverse shear strains γ_{yz} =0 and similarly γ_{xz} =0. However, γ_{xy} does not equal 0, right angles in the plane of the plate may not remain right angles after loading. The plate may twist in the plane.

- Thickness changes can be neglected and normals undergo no extension. This means normal strain, ε_z=0.

- Normal stresses σz has no effect on in-plane strains εx and εy in the stress-strain equations and is considered negligible.

- Membrane or in-plane forces are neglected here and the plane stress resistance can be superimposed later. That is, the in-plane deformations in the x and y directions at the mid surface are assumed to be zero, u(x,y,0)=0 and v(x,y,0)=0.

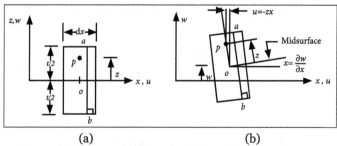

(a)   (b)

(1) Differential slice of plate of thickness t(a) before loading and (b) displacements of point P after loading, based on Kirchhoff theory.

Transverse shear deformation is neglected and so right angles in the cross section remain right angles. Displacements in the y-z plane are similar.

Based on the Kirchhoff assumptions, any point P in the figure (1) has displacement in the x direction due to a small rotation α of:

$$u = -z\alpha = -z\left(\frac{\partial w}{\partial x}\right) \qquad \qquad \dots(1)$$

and similarly the same point has displacement in the y direction of:

$$v = -z\left(\frac{\partial w}{\partial x}\right) \qquad \qquad \dots(2)$$

The curvatures of the plate are then given as the rate of change of the angular displacements of the normals and are defined as:

$$k_x = -\frac{\partial^2 w}{\partial x^2} \quad k_y = -\frac{\partial^2 w}{\partial y^2} \quad k_{xy} = -\frac{2\partial^2 w}{\partial x \partial y} \qquad \dots(3)$$

The first of equations (3) is used in beam theory:

$$\varepsilon_x = \frac{\partial u}{\partial x} \qquad \varepsilon_y = \frac{\partial v}{\partial x} \qquad \gamma_{xy} = \frac{\partial u}{\partial y} + \frac{\partial v}{\partial x} \qquad \dots(a)$$

Using the definitions for the in-plane strains from equation (a), along with equation (3), the in-plane strain/displacement equations become:

$$\varepsilon_x = -z\frac{\partial^2 w}{\partial x^2} \quad \varepsilon_y = -z\frac{\partial^2 w}{\partial y^2} \quad \gamma_{xy} = -2z\frac{\partial^2 w}{\partial x \partial y} \qquad \dots(4a)$$

or using equation (3) in equation(4a), we have:

$$\varepsilon_x = -zk_x \qquad \varepsilon_y = -zk_y \qquad \gamma_{xy} = -zk_{xy} \qquad \dots(4b)$$

The first of equations (4a) is used in beam theory. The others are new to plate theory.

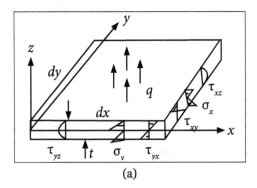

(a)

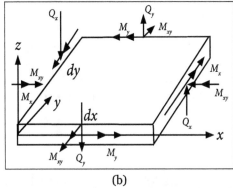

(b)

(2)Differential element of a plate with (a) stresses shown on the edges of the plate and (b) differential moments and forces.

Stress/Strain Relations

Based on the third assumption above, the plane stress equations can be used to relate the in-plane stresses to the in-plane strains for an isotropic material as:

$$\sigma_x = \frac{E}{1-v^2}\left(\varepsilon_x - v\varepsilon_y\right)$$

$$\sigma_y = \frac{E}{1-v^2}\left(\varepsilon_y - v\varepsilon_x\right)$$

$$\gamma_{xy} = G\gamma_{xy} \qquad\qquad ...(5)$$

The in-plane normal stresses and shear stress are shown acting on the edges of the plate in the figure 2(a). Similar to the stress variation in a beam, these stresses vary linearly in the z direction from the mid-surface of the plate.

The transverse shear stresses τ_{yz} and τ_{xz} are also present, even though transverse shear deformation is neglected. As in beam theory, these transverse stresses vary quadratically through the plate thickness.

The stresses of equation (5) can be related to the bending moments M_x and M_y and to the twisting moment M_{xy} acting along the edges of the plate as shown in the figure 2(b).

The moments are actually functions of x and y and are computed per unit length in the plane of the plate. Therefore, the moments are:

$$M_x = \int_{-t/2}^{t/2} z\sigma_x\, dz \qquad M_y = \int_{-t/2}^{t/2} z\sigma_y\, dz \qquad M_{xy} = \int_{-t/2}^{t/2} z\tau_{xy}\, dz \qquad ...(6)$$

The moments can be related to the curvatures by substituting equations (4b) into equations (5) and then using those stresses in equation (6) to obtain:

$$M_x = D\left(\kappa_x + v\kappa_y\right) \qquad M_y = D\left(\kappa_y + v\kappa_x\right) \qquad M_{xy} = \frac{D(1-v)}{2}\kappa_{xy} \qquad ...(7)$$

Where, $D = Et^3 / [12(1-v^2)]$ is called the bending rigidity of the plate.

The maximum magnitudes of the normal stresses on each edge of the plate are located at the top or bottom at z=t/2. For instance, it can be shown that:

$$\sigma_x = \frac{6Mx}{t^2} \qquad\qquad ...(8)$$

This formula is similar to the flexure formula $\sigma_x = M_x c/1$ when applied to a unit width of plate and when c=t/2.

The governing equilibrium differential equation of plate bending is important in selecting the element displacement fields. The basis for this relationship is the equilibrium differential equations derived by the equilibrium of forces with respect to the z direction and by the equilibrium of moments about the x and y axes, respectively.

These equilibrium equations result in the following differential equations:

$$\frac{\partial Q_x}{\partial x} + \frac{\partial Q_y}{\partial y} + q = 0$$

$$\frac{\partial M_x}{\partial x} + \frac{\partial M_{xy}}{\partial y} - Q_x = 0$$

$$\frac{\partial M_y}{\partial y} + \frac{\partial M_{xy}}{\partial x} - Q_y = 0 \qquad \qquad ...(9)$$

Where, q is the transverse distributed loading and Qx and Qy are the transverse shear line loads shown in the figure 2(b).

Now substituting the moment/curvature relations from equation (7) into the second and third of equations (9), then solving those equations for Q_x and Q_y and finally substituting the resulting expressions into the first of equations (9), we obtain the governing partial differential equation for an isotropic, thin-plate bending behavior as:

$$D\left(\frac{\partial^4 w}{\partial x^4} + \frac{2\partial^4 w}{\partial x^2 \partial y^2} + \frac{\partial^4 w}{\partial y^4}\right) = q \qquad \qquad ...(10)$$

From equation (10), we observe that the solution of thin-plate bending using a displacement point of view depends on selection of the single-displacement component w, the transverse displacement.

Potential Energy of a Plate

The total potential energy of a plate is given by:

$$U = \frac{1}{2}\int \left(\sigma_x \varepsilon_x + \sigma_y \varepsilon_y + \tau_{xy} \gamma_{xy}\right) dV \qquad \qquad ...(11)$$

The potential energy can be expressed in terms of the moments and curvatures by substituting equations (4b) and (6) in equation (11) as:

$$U = \frac{1}{2}\int \left(M_x \kappa_x + M_y \kappa_y + M_{xy} \kappa_{xy}\right) dA \qquad \qquad ...(12)$$

4.8   Finite Element Analysis of Shell

A shell is a curved surface, which by virtue of their shape can withstand both membrane and bending forces. A shell structure can take higher loads if, membrane stresses are predominant, which is primarily caused due to in-plane forces (plane stress condition).

However, localized bending stresses will appear near load concentrations or geometric discontinuities. The shells are analogous to cable or arch structure depending on whether the shell resists tensile or compressive stresses respectively.

Few advantages using shell elements are given below:

- Higher load carrying capacity.

- Lesser thickness and hence lesser dead load.

- Lesser support requirement.

- Larger useful space.

- Higher aesthetic value.

The example of shell structures includes large-span roof, cooling towers, piping system, pressure vessel, aircraft fuselage, rockets, water tank, arch dams and many more. Even in the field of biomechanics, shell elements are used for analysis of skull, Crustaceans shape, red blood cells, etc.

Classification of Shells

Shell may be classified with several alternatives. Depending upon deflection in transverse direction due to transverse shear force per unit length, the shell can be classified into structurally thin or thick shell.

Further, depending upon the thickness of the shell in comparison to the radii of curvature of the mid surface, the shell is referred to as geometrically thin or thick shell.

Typically, if thickness to radii of curvature is less than 0.05, then the shell can be assumed as a thin shell. For most of the engineering application the thickness of shell remains within 0.001 to 0.05 and treated as thin shell.

Assumptions for Thin Shell Theory

Thin shell theories are basically based on Love-Kirchhoff assumptions as follows:

- As the shell deforms, the normal to the un-deformed middle surface remain straight and normal to the deformed middle surface undergo no extension. i.e., all strain components in the direction of the normal to the middle surface is zero.

- The transverse normal stress is neglected.

Thus, above assumptions reduce the three dimensional problems into two dimensional.

Overview of Shell Finite Elements

Many approaches exist for deriving shell finite elements, such as, flat shell element, curved shell element, solid shell element and degenerated shell element.

(A) Flat Shell Element

The geometry of these types of elements is assumed as flat. The curved geometry of shell is obtained by assembling number of flat elements. These elements are based on combination of membrane element and bending element that enforced Kirchhoff's hypothesis.

It is important to note that the coupling of membrane and bending effects due to curvature of the shell is absent in the interior of the individual elements.

(B) Curved Shell Element

Curved shell elements are symmetrical about an axis of rotation. As in case of axisymmetric plate elements, membrane forces for these elements are represented with respected to meridian direction as (u, Nz, M_θ) and in circumferential directions as (w, N_θ, M_z).

However, the difficulties associated with these elements includes, difficulty in describing geometry and achieving inter-elemental compatibility. Also, the satisfaction of rigid body modes of behavior is acute in curved shell elements.

(C) Solid Shell Element

Though, use of 3D solid element is another option for analysis of shell structure, dealing with too many degrees of freedom makes it uneconomic in terms of computation time.

Further, due to small thickness of shell element, the strain normal to the mid surface is associated with very large stiffness coefficients and thus makes the equations ill conditioned.

(D) Degenerated Shell Elements

Here, elements are derived by degenerating a 3D solid element into a shell surface element, by deleting the intermediate nodes in the thickness direction and then by projecting the nodes on each surface to the mid surface as shown in the figure(1).

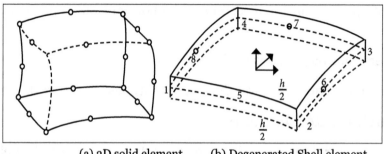

(a) 3D solid element   (b) Degenerated Shell element.
(1) Degeneration of 3D element.

This approach has the advantage of being independent of any particular shell theory. This approach can be used to formulate a general shell element for geometric and material nonlinear analysis.

Such element has been employed very successfully when used with 9 or in particular, 16 nodes. However, the 16-node element is quite expensive in computation.

In a degenerated shell model, the numbers of unknowns present are five per node (three mid-surface displacements and two director rotations). Moderately thick shells can be analyzed using such elements.

However, selective and reduced integration techniques are necessary to use due to shear locking effects in case of thin shells. The assumptions for degenerated shell are similar to the Reissner Minding assumptions.

Finite Element Formulation of a Degenerated Shell

Let consider a degenerated shell element, obtained by degenerating 3D solid element. The degenerated shell element as shown in the figure 1(b) has eight nodes, for which the analysis is carried out.

Let (ξ, η) are the natural coordinates in the mid surface. And ς is the natural coordinate along thickness direction.

The shape functions of a two dimensional eight node Iso parametric element are:

$$N_1 = \frac{(1-\xi)(1-\eta)(-\xi-\eta-1)}{4} \qquad N_5 = \frac{(1+\xi)(1-\eta)(1-\eta)}{2}$$

$$N_2 = \frac{(1+\xi)(1-\eta)(\xi-\eta-1)}{4} \qquad N_6 = \frac{(1+\xi)(1+\eta)(1-\eta)}{2}$$

$$N_3 = \frac{(1+\xi)(1+\eta)(\xi+\eta-1)}{4} \qquad N_7 = \frac{(1+\xi)(1-\xi)(1+\eta)}{2}$$

$$N_4 = \frac{(1-\xi)(1+\eta)(-\xi+\eta-1)}{4} \qquad N_8 = \frac{(1-\xi)(1+\eta)(1-\eta)}{2} \qquad \text{...(1)}$$

The position of any point inside the shell element can be written in terms of nodal coordinates as:

$$\begin{Bmatrix} x \\ y \\ z \end{Bmatrix} = \sum_{i=1}^{8} N_i(\xi,\eta) \left\{ \frac{1+\varsigma}{2} \begin{Bmatrix} x_i \\ y_i \\ z_i \end{Bmatrix}_{top} + \frac{1-\varsigma}{2} \begin{Bmatrix} x_i \\ y_i \\ z_i \end{Bmatrix}_{bottom} \right\} \qquad ...(2)$$

Since, ς is assumed to be normal to the mid surface, the above expression can be rewritten in terms of a vector connecting the upper and lower points of shell as:

$$\begin{Bmatrix} x \\ y \\ z \end{Bmatrix} = \sum_{i=1}^{8} N_i(\xi,\eta) \left\{ \frac{1}{2} \left\{ \begin{Bmatrix} x_i \\ y_i \\ z_i \end{Bmatrix}_{top} + \begin{Bmatrix} x_i \\ y_i \\ z_i \end{Bmatrix}_{bottom} \right\} + \frac{\varsigma}{2} \left\{ \begin{Bmatrix} x_i \\ y_i \\ z_i \end{Bmatrix}_{top} - \begin{Bmatrix} x_i \\ y_i \\ z_i \end{Bmatrix}_{bottom} \right\} \right\}$$

Or

$$\begin{Bmatrix} x \\ y \\ z \end{Bmatrix} = \sum_{i=1}^{8} N_i(\xi,\eta) \left\{ \begin{Bmatrix} x_i \\ y_i \\ z_i \end{Bmatrix} + \frac{\varsigma}{2} V_{3i} \right\} \qquad ...(3)$$

Where,

$$\begin{Bmatrix} x_i \\ y_i \\ z_i \end{Bmatrix} = \frac{1}{2} \left\{ \begin{Bmatrix} x_i \\ y_i \\ z_i \end{Bmatrix}_{top} + \begin{Bmatrix} x_i \\ y_i \\ z_i \end{Bmatrix}_{bottom} \right\} \text{ and, } V_{3i} = \begin{Bmatrix} x_i \\ y_i \\ z_i \end{Bmatrix}_{top} - \begin{Bmatrix} x_i \\ y_i \\ z_i \end{Bmatrix}_{bottom} \qquad ...(4)$$

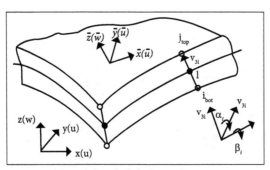

(a) Local and global coordinates.

For small thickness, the vector V_{3i} can be represented as a unit vector tiv3i:

$$\begin{Bmatrix} x \\ y \\ z \end{Bmatrix} = \sum_{i=1}^{8} N_i(\xi,\eta) \left\{ \begin{Bmatrix} x_i \\ y_i \\ z_i \end{Bmatrix} + \frac{\varsigma}{2} t_i v_{3i} \right\} \qquad ...(5)$$

Where, t_i is the thickness of shell at i_{th} node. In a similar way, the displacement at any point of the shell element can be expressed in terms of three displacements and two rotation components about two orthogonal directions normal to nodal load vector V_{3i} as:

$$\begin{Bmatrix} u \\ v \\ w \end{Bmatrix} = \sum_{i=1}^{8} N_i(\xi,\eta) \left\{ \begin{Bmatrix} u_i \\ v_i \\ w_i \end{Bmatrix} + \frac{\varsigma t_i}{2} \begin{bmatrix} v_{1i} - v_{2i} \end{bmatrix} \begin{Bmatrix} \alpha_i \\ \beta_i \end{Bmatrix} \right\} \qquad ...(6)$$

Where, (α_i, β_i) are the rotations of two unit vectors v_{1i} & v_{2i} about two orthogonal directions normal to nodal load vector V_{3i}. The values of v_{1i} and v_{2i} can be calculated in following way.

The coordinate vector of the point to which a normal direction is to be constructed may be defined as:

$$x = x\hat{i} + y\hat{j} + z\hat{k} \qquad ...(7)$$

In which, $\hat{i}, \hat{j}, \hat{k}$ are three (orthogonal) base vectors. Then, Vu is the cross product of $\hat{i}$ & V_{3i} as shown below:

$$V_{1i} = \hat{i} \times V_{3i} \quad \& \quad V_{2i} = V_{3i} \times V_{1i} \qquad ...(8)$$

And

$$v_{1i} = {V_{1i}}\big/{|V_{1i}|} \quad \& \quad v_{2i} = {V_{2i}}\big/{|V_{2i}|} \qquad ...(9)$$

Jacobin Matrix

The Jacobian matrix for eight node shell element can be expressed as:

$$[J] = \begin{bmatrix} \sum_{i=1}^{8}\left(x_i + tx'_i\right)\dfrac{\partial N_i}{\partial \xi} & \sum_{i=1}^{8}\left(y_i + ty'_i\right)\dfrac{\partial N_i}{\partial \xi} & \sum_{i=1}^{8}\left(z_i + tz'_i\right)\dfrac{\partial N_i}{\partial \xi} \\ \sum_{i=1}^{8}\left(x_i + tx'_i\right)\dfrac{\partial N_i}{\partial \eta} & \sum_{i=1}^{8}\left(y_i + ty'_i\right)\dfrac{\partial N_i}{\partial \eta} & \sum_{i=1}^{8}\left(z_i + tz'_i\right)\dfrac{\partial N_i}{\partial \eta} \\ \sum_{i=1}^{8} Nx'_i & \sum_{i=1}^{8} Ny'_i & \sum_{i=1}^{8} Nz'_i \end{bmatrix} \qquad ...(10)$$

Strain Displacement Matrix

The relationship between strain and displacement is described by:

$$\{\varepsilon\} = [B]\ \{d\} \qquad ...(11)$$

Where, the displacement vector will become:

$$\{d\}^T = \{u_1 v_1 w_1 v_{1i} v_{2i} \ldots\ldots u_8 v_8 w_8 v_{18} v_{28}\}$$

And the strain components will be:

$$[\varepsilon] = \begin{Bmatrix} \dfrac{\partial u}{\partial x} \\[2mm] \dfrac{\partial v}{\partial y} \\[2mm] \dfrac{\partial u}{\partial y} + \dfrac{\partial v}{\partial x} \\[2mm] \dfrac{\partial v}{\partial y} + \dfrac{\partial w}{\partial y} \\[2mm] \dfrac{\partial w}{\partial x} + \dfrac{\partial u}{\partial z} \end{Bmatrix} \qquad \ldots(13)$$

Using equation (6) in equation (13) and then differentiating with respect to (ξ, η, ς) the strain displacement matrix will be obtained as:

$$\begin{bmatrix} \dfrac{\partial u}{\partial \xi} & \dfrac{\partial v}{\partial \xi} & \dfrac{\partial w}{\partial \xi} \\[2mm] \dfrac{\partial u}{\partial \eta} & \dfrac{\partial v}{\partial \eta} & \dfrac{\partial w}{\partial \eta} \\[2mm] \dfrac{\partial u}{\partial \varsigma} & \dfrac{\partial v}{\partial \varsigma} & \dfrac{\partial w}{\partial \varsigma} \end{bmatrix} = \sum_{i=1}^{8} \begin{bmatrix} \dfrac{\partial N_i}{\partial \xi} \\[2mm] \dfrac{\partial N_i}{\partial \eta} \\[2mm] 0 \end{bmatrix} \begin{bmatrix} u_i & v_i & w_i \end{bmatrix} - \sum_{i-1}^{8} \dfrac{t_1 v_{2i}}{2}$$

$$\begin{bmatrix} \varsigma \dfrac{\partial N_i}{\partial \xi} \\[2mm] \varsigma \dfrac{\partial N_i}{\partial \eta} \\[2mm] N_i \end{bmatrix} \times \begin{bmatrix} \beta_1 \\ \beta_2 \\ \beta_3 \end{bmatrix}^T + \sum_{i-1}^{8} \dfrac{t_1 v_{1i}}{2} \begin{bmatrix} \varsigma \dfrac{\partial N_i}{\partial \xi} \\[2mm] \varsigma \dfrac{\partial N_i}{\partial \eta} \\[2mm] N_i \end{bmatrix} \times \begin{bmatrix} \alpha_1 \\ \alpha_2 \\ \alpha_3 \end{bmatrix}^T \qquad \ldots(14)$$

Stress Strain Relation

The stress strain relationship is given by:

$$\{\sigma\} = [D]\{\varepsilon\} \qquad \ldots(15)$$

Using equation (11) in equation (15) one can find the following relation:

$$\{\sigma\} = [D][B]\{d\} \qquad \qquad ...(16)$$

Where, the stress strain relationship matrix is represented by:

$$[D] = \frac{E}{1-\mu^2} \begin{bmatrix} 1 & \mu & 0 & 0 & 0 \\ \mu & 1 & 0 & 0 & 0 \\ 0 & 0 & \dfrac{1-\mu}{2} & 0 & 0 \\ 0 & 0 & 0 & \dfrac{\alpha(1-\mu)}{2} & 0 \\ 0 & 0 & 0 & 0 & \dfrac{\alpha(1-\mu)}{2} \end{bmatrix} \qquad ...(17)$$

The value of shear correction factor α is considered generally as 5/6. The above constitutive matrix can be split into two parts ($[D_b]$ and $[D_s]$ for adoption of different numerical integration schemes for bending and shear contributions to the stiffness matrix.

$$[D] = \begin{bmatrix} [D_b] & \vdots & [0] \\ \cdots & \cdots & \cdots \\ [0] & \vdots & [D_s] \end{bmatrix} \qquad ...(18)$$

Thus,

$$[D_b] = \frac{E}{1-\mu^2} \begin{bmatrix} 1 & \mu & 0 \\ \mu & 1 & 0 \\ 0 & 0 & \dfrac{1-\mu}{2} \end{bmatrix} \qquad ...(19)$$

And $\quad [D_s] = \dfrac{E\alpha}{2(1+\mu)} \begin{bmatrix} 1 & 0 \\ 0 & 1 \end{bmatrix} \qquad ...(20)$

It may be important to note that the constitutive relation expressed in equation (19) is same as for the case of plane stress formulation. Also, equation (20) with a multiplication of thickness h is similar to the terms corresponds to shear force in case of plate bending problem.

Element Stiffness Matrix

Finally, the stiffness matrix for the shell element can be computed from the expression:

$$[k] = \iiint [B]T\, [D]\, [B]\, d\Omega \qquad ...(21)$$

However, it is convenient to divide the elemental stiffness matrix into two parts: (i) bending and membrane effect and (ii) transverse shear effects. This will facilitate the use of appropriate order of numerical integration of each part. Thus:

$$[k] = [k]_b + [k]_s \qquad \qquad \ldots(22)$$

Where, contribution due to bending and membrane effects to stiffness is denoted as [k]b and transverse shear contribution to stiffness is denoted as [k]s and expressed in the following form:

$$[k]_b = \iiint [B]_b^T [D]_b [B]_b \, d\Omega \text{ and } [k]_s = \iiint [B]_s^T [D]_s [B]_s \, d\Omega \qquad \ldots(23)$$

Numerical procedure will be used to evaluate the stiffness matrix. A 2 ×2 Gauss Quadrature can be used to evaluate the integral of [k]b and one point Gauss Quadrature may be used to integrate [k]s to avoid shear locking effect.

4.9 Additional Applications of FEM: Finite Elements for Elastic Stability

There are two types of structural failure associated, namely (i) material failure and (ii) geometry or configuration failure. In case of material failure, the stresses exceed the permissible values which may lead to formation of cracks.

In configuration failure, the structure is unable to maintain its designed configuration under the external disturbance even though the stresses are in permissible range.

The stability loss due to tensile forces falls in broad category of material instability. The loss of stability under compressive load is termed as structural or geometric instability which is commonly known as buckling.

Thus, buckling is considered by a sudden failure of a structural element subjected to large compressive stress, where the actual compressive stress at the point of failure is less than the ultimate allowable compressive stresses. Buckling is a wide term that describes a range of mechanical behaviour.

Generally, it refers to an incident where a structural element in compression deviates from a behaviour of elastic shortening within the original geometry and undergoes large deformations involving a change in member shape for a small increase in force. Various types of buckling may occur such as flexural buckling, torsional buckling, torsional flexural buckling etc.

Buckling of Truss Members

For a truss member, the axial strain can be expressed in terms of its displacements:

$$\varepsilon_x = \frac{\partial u}{\partial x} + \frac{1}{2}\left[\left(\frac{\partial u}{\partial x}\right)^2 + \left(\frac{\partial v}{\partial x}\right)^2\right] \qquad ...(1)$$

Here, u and v are the displacements in local X and Y directions respectively. Here, the strain-displacement relation is nonlinear. The total potential energy in the member with uniform cross sectional area and subjected to external forces can be written as:

$$\Pi = \frac{EA}{2}\int_0^L \varepsilon_x{}^2 dx - \left(P_1 v_1 + P_2 v_2\right) \qquad ...(2)$$

Where, P_1 and P_2 are the external forces in nodes 1 and 2, $v1$ and $v2$ are the vertical displacements in nodes 1 and 2, E and A are the modulus of elasticity and cross sectional area respectively (Figure a). The length of the member is considered to be L.

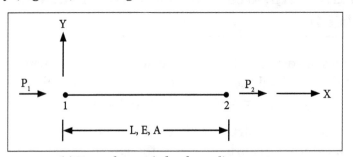

(a) Truss element in local coordinate system.

The above expression can be rewritten in terms of displacement variables as:

$$\Pi = \frac{EA}{2}\int_0^L \left(\frac{\partial u}{\partial x} + \frac{1}{2}\left[\left(\frac{\partial u}{\partial x}\right)^2 + \left(\frac{\partial v}{\partial x}\right)^2\right]\right) dx - \left(P_1 v_1 + P_2 v_2\right)$$

$$= \frac{EA}{2}\int_0^L \left\{\left[\left(\frac{\partial u}{\partial x}\right)^2 + \left(\frac{\partial u}{\partial x}\right)\left(\frac{\partial v}{\partial x}\right)^2 + \left(\frac{\partial u}{\partial x}\right)^3\right] + \frac{1}{4}\left[\left(\frac{\partial u}{\partial x}\right)^2\right.\right.$$

$$\left.\left. + \left(\frac{\partial v}{\partial x}\right)^2\right]^2\right\} dx - \left(P_1 v_1 + P_2 v_2\right) \qquad ...(3)$$

Neglecting higher order terms and considering $EA\dfrac{\partial u}{\partial x}=P$ the above expression can simplify to the following form:

$$\Pi=\frac{EA}{2}\int_0^L\left(\frac{\partial u}{\partial x}\right)^2 dx+\frac{P}{2}\int_0^L\left(\frac{\partial v}{\partial x}\right)^2 dx-\left(P_1v_1+P_2v_2\right) \qquad ...(4)$$

Considering nodal displacements at nodes 1 and 2 as u1, v1 and u2, v2, the displacement at any point inside the element can be represented in terms of its interpolation functions and nodal displacements.

The shape function for a two node truss element is:

$$[N]=\left[1-\frac{x}{L} \quad \frac{x}{L}\right] \qquad ...(5)$$

Equation expressed in equation (5) is expressed in terms of the nodal displacement vectors $\{u_i\}$ and $\{v_i\}$ with the help of interpolation function.

$$\Pi=\frac{1}{2}\int_0^L\{u_1\}^T[N']^T EA[N']\{u_1\}dx+\frac{P}{2}\int_0^L\{v_i\}^T[N']^T[N']\{v_i\}dx-\left(P_1v_1+P_2v_2\right) \quad ...(6)$$

Where,

$$\frac{\partial u}{\partial x}=\frac{\partial}{\partial x}\left([N]\{u_1\}\right)=[N']\{u_i\} \text{ and } \frac{\partial v}{\partial x}=\frac{\partial}{\partial x}\left([N]\{v_i\}\right)=[N']\{v_i\} \qquad ...(7)$$

Making potential energy (equation 6) stationary one can find the equilibrium equation as:

$$\{F\}=\int_0^L[N']^T EA[N']dx\{u_i\}+P\int_0^L[N']^T[N']dx\{v_i\} \qquad ...(8)$$

$$\text{Or } \{F\}=[k_A]\{u_i\}+P[k_G]\{v_i\} \qquad ...(9)$$

Where, $[k_A]$ is the axial stiffness of the member and $[k_G]$ is the geometric stiffness of the member in its local coordinate system and can be derived as follows:

$$[k_A]=\int_0^L[N']^T EA[N']dx=AE\int_0^L\begin{Bmatrix}-\dfrac{1}{L}\\ \dfrac{1}{L}\end{Bmatrix}\begin{bmatrix}-\dfrac{1}{L} & \dfrac{1}{L}\end{bmatrix}dx=\frac{AE}{L}\begin{bmatrix}1 & -1\\ -1 & 1\end{bmatrix} \qquad ...(10)$$

$$[k_G] = \int_0^L [N']^T [N'] dx = \int_0^L \left\{ \begin{array}{c} -\dfrac{1}{L} \\ \dfrac{1}{L} \end{array} \right\} \left[-\dfrac{1}{L} \quad \dfrac{1}{L} \right] dx = \dfrac{1}{L} \begin{bmatrix} 1 & -1 \\ -1 & 1 \end{bmatrix} \qquad ...(11)$$

However, in a generalized form to accommodate both the direction, these stiffness matrices can be written as:

$$[k_A] = \dfrac{AE}{L} \begin{bmatrix} 1 & 0 & -1 & 0 \\ 0 & 0 & 0 & 0 \\ -1 & 0 & 1 & 0 \\ 0 & 0 & 0 & 0 \end{bmatrix} \text{ and } [k_G] = \dfrac{1}{L} \begin{bmatrix} 0 & 0 & 0 & 0 \\ 0 & 1 & 0 & -1 \\ 0 & 0 & 0 & 0 \\ 0 & -1 & 0 & 1 \end{bmatrix} \qquad ...(12)$$

The generalized stiffness matrix of a plane truss member in global coordinate system can be derived using the transformation matrix. The transformation matrix can be written as follows:

$$[T] = \begin{bmatrix} \cos\theta & \sin\theta & 0 & 0 \\ -\sin\theta & \cos\theta & 0 & 0 \\ 0 & 0 & \cos\theta & \sin\theta \\ 0 & 0 & -\sin\theta & \cos\theta \end{bmatrix} \qquad ...(13)$$

Thus, the stiffness matrices with respect to global coordinate system will become:

$$\left[K_A \right] = [T]^T \left[k_A \right] [T] \text{ and } \left[K_G \right] = [T]^T \left[k_G \right] [T] \qquad ...(14)$$

Here, [KA] and [KG] are the axial stiffness and geometric stiffness of the member in global coordinate system. The force-displacement relationship in global coordinate system can be written from equation (9) as:

$$\{F\} = \left[K_A \right]\{d\} + P\left[K_G \right]\{d\} = \left[\left[K_A \right] + P[K_G] \right]\{d\} \qquad ...(15)$$

Where, {d} is the displacement vector in global coordinates. If the external force is absent in equation (15), the value of P will be considered to be undetermined as:

$$\left[K_A \right]\{d\} + P\left[K_G \right]\{d\} = 0 \qquad ...(16)$$

The above equation can be solved as an eigenvalue problem to calculate buckling load P.

Buckling of Beam-Column Members

Let consider a pin ended column under the action of compressive force P. The elastic and geometric stiffness matrices can be developed from 1st principles for a beam-column element which can be used in the linear elastic stability analysis of frameworks.

Considering small deflection approximation to the curvature, the total potential energy is computed from the following:

$$\Pi=\frac{EI}{2}\int_0^L\left(\frac{d^2w}{dx^2}\right)^2dx+\frac{P}{2}\int_0^L\left(\frac{dw}{dx}\right)^2dx \quad =\int_0^L\left[\frac{EI}{2}\left(\frac{d^2w}{dx^2}\right)^2+\frac{P}{2}\left(\frac{dw}{dx}\right)^2\right]dx \quad ...(17)$$

Here, w is the transverse displacement and I is the moment of inertia of the member.

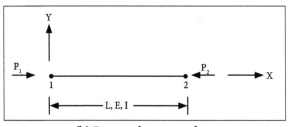

(b) Beam-column member.

Considering nodal displacements at nodes 1 and 2 as w1, θ1 and w2, θ2, the displacement at any point inside the element can be represented in terms of its interpolation functions and nodal displacements.

$$w=\begin{bmatrix}N_1 & N_2 & N_3 & N_4\end{bmatrix}\begin{bmatrix}w_1\\\theta_1\\w_2\\\theta_2\end{bmatrix}=[N]\{d\} \quad ...(18)$$

Thus, the above energy equation can be rewritten as:

$$\Pi=\frac{1}{2}\int_0^L[\{d\}]^T[N'']^T EI[N'']\{d\}+\{d\}P[N']^T[N']\{d\}dx \quad ...(19)$$

Where,

$$\frac{d^2w}{dx^2}=\frac{d^2}{dx^2}([N]\{d\})=[N'']\{d\} \text{ and } \frac{dw}{dx}=\frac{d}{dx}([N]\{d\})=[N']\{d\} \quad ...(20)$$

Applying the variation principle one can express:

$$\{F\}=\frac{\partial\Pi}{\partial\{d\}}=\int_0^1\left([N'']^T EI[N'']+P[N']^T[N']dx\{d\}\right) \quad ...(21)$$

Thus, the stiffness matrix will be obtained as follows which have two terms:

$$\{k\}=\int_0^1\left([N'']^T EI[N'']\right)dx+P\int_0^1\left([N']^T[N']dx\right)=[k_F]+P[k_G] \quad ...(22)$$

The first term resembles ordinary stiffness matrix for the bending of a beam. So this matrix is called flexural stiffness matrix. The second matrix is known as geometric stiffness matrix as it only depends on the geometrical parameters. Thus, the flexural stiffness matrix $[k_F]$ and geometric stiffness matrix $[k_G]$ can be derived from the following expressions.

$$[k_F] = \int_0^1 \left([N'']^T EI [N'']\right) dx \quad \text{and} \quad [k_G] = \int_0^1 \left([N']^T [N'] dx\right) \qquad ...(23)$$

The above matrices can be derived from the assumed interpolation function. For example, the interpolation functions for a two node beam element are expressed by the following equation:

$$[N] = \left[\left(1 - 3\frac{x^2}{L^2} + 2\frac{x^3}{L^3}\right), \left(x - 2\frac{x^2}{L} + \frac{x^3}{L^2}\right), \left(3\frac{x^2}{L^2} - 2\frac{x^3}{L^3}\right), \left(-\frac{x^2}{L} + \frac{x^3}{L^2}\right) \right] \qquad ...(24)$$

Now, the first and second order derivative of the above function will become:

$$[N'] = \left[\left(-6\frac{x}{L^2} + 6\frac{x^2}{L^3}\right), \left(1 - 4\frac{x}{L} + 3\frac{x^2}{L^2}\right), \left(6\frac{x}{L^2} - 6\frac{x^2}{L^3}\right), \left(-2\frac{x}{L} + 3\frac{x^2}{L^2}\right) \right] \qquad ...(25)$$

$$[N''] = \left[\left(-\frac{6}{L^2} + 12\frac{x}{L^3}\right), \left(-\frac{4}{L} + 6\frac{x}{L^2}\right), \left(\frac{6}{L^2} - 12\frac{x}{L^3}\right), \left(-\frac{2}{L} + 6\frac{x}{L^2}\right) \right] \qquad ...(26)$$

Using the above expressions in equation (23), flexural and geometric stiffness matrices can be derived and obtained as follows:

$$[K_F] = \frac{EI}{L^3} \begin{bmatrix} 12 & 6L & -12 & 6L \\ 6L & 4L^2 & -6L & 2L^2 \\ -12 & -6L & 12 & -6L \\ 6L & 2L^2 & -6L & 4L^2 \end{bmatrix} \qquad ...(27)$$

$$[K_G] = \frac{1}{30L} \begin{bmatrix} 36 & 3L & -36 & 3L \\ 3L & 4L^2 & -3L & -L^2 \\ 36 & -3L & 36 & -3L \\ 3L & -L^2 & -3L & 4L^2 \end{bmatrix} \qquad ...(28)$$

Due to external forces $\{F\}$, one can find the displacement vectors from the following equation:

$$\{F\} = [k] \{d\} = [k_F]\{d\} + P[k_G]\{d\} \qquad ...(29)$$

The above is the beam-column equation in finite element form. In the absence of transverse load, the member will become column and P will be considered to be undetermined:

$$\left[k_F\right]\{d\} + P\left[k_G\right]\{d\} = 0 \qquad \qquad ...(30)$$

Denoting P as -λ we can reach to the familiar eigenvalue problem as given below:

$$\left[k_F\right]\{d\} = \lambda\left[k_G\right]\{d\} \qquad \qquad ...(31)$$

Solving above equation, values of λ and associated nodal displacement vectors can be obtained.

Mathematically this can be expressed as:

$$\left(\left[k_F\right] - \lambda\left[k_G\right]\right)\{d\} = 0 \qquad \qquad ...(32)$$

Thus,

$$\left\|\left[k_F\right] - \lambda\left[k_G\right]\right\| \qquad \qquad ...(33)$$

For the matrix of size n, one can find n+1 degree polynomial in λ. The smallest root of the above equation will become the first approximate buckling load. From this value λ, one can find a set of ratios for the nodal displacement components.

From this, the first buckling mode shape can be calculated. Higher mode approximations can also be found in a similar process. The procedure to determine the critical load by the above method is illustrated in the following example.

Problem:

1. Consider a column with one end clamped and other end free as shown in the figure (a). Let us determine the critical load.

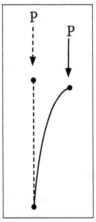

(a) Column with one end fixed and other end free.

Solution

Given:

Now, the finite element equation of the column considering single element will be:

$$
\frac{EI}{L^3}\begin{bmatrix} 12 & 6L & -12 & 6L \\ 6L & 4L^2 & -6L & 2L^2 \\ -12 & -6L & 12 & -6L \\ 6L & 2L^2 & -6L & 4L^2 \end{bmatrix} - \lambda \frac{1}{30L}\begin{bmatrix} 36 & 3L & -36 & 3L \\ 3L & 4L^2 & -3L & -L^2 \\ 36 & -3L & 36 & -3L \\ 3L & -L^2 & -3L & 4L^2 \end{bmatrix} = 0
$$

The boundary conditions for this member are given by:

$$
\text{At } x=0,\ d_1=0 \text{ and } \theta_1=\frac{dw}{dx}=0
$$

Thus, according to the above boundary conditions, the first and second rows as well as columns of the equation are deleted and rewritten in the following form:

$$
\left| \frac{EI}{L^3}\begin{bmatrix} 12 & -6L \\ -6L & 4L^2 \end{bmatrix} - \frac{\lambda}{30L}\begin{bmatrix} 36 & -3L \\ -3L & 4L^2 \end{bmatrix} \right| = 0
$$

Or

$$
\left(\frac{4EI}{L}-\frac{2\lambda L}{15}\right)\left(\frac{12EI}{L^3}-\frac{6\lambda}{5L}\right)-\left(-\frac{6EI}{L^2}+\frac{\lambda}{10}\right)^2=0
$$

Solving the above expression, the critical value of λ_{cr} and thus the critical value of force Pcr will become as:

$$
\lambda_{cr}=P_{cr}=2.486\frac{EI}{L^2}
$$

It is important to note that the exact value for such clamped-free column is given by:

$$
P_{cr}=\frac{\pi^2 EI}{L_e^2}=\frac{\pi^2 EI}{(2L)^2}=2.467\frac{EI}{L^2}
$$

The finite element result has slight deviation from the exact result. This difference can be minimized by the increase of number of elements in the column as we know, more we subdivide the continuum, better we can obtain the result close to the exact one.

Buckling of Plate Bending Elements

The elastic stability analysis of rectangular plates is shown below. The total potential energy for plate are expressed as:

$$\pi = \frac{D}{2} \int_{-a}^{a} \int_{-b}^{b} \left\{ \left(\nabla^2 w \right)^2 + 2(1-\mu) \left[\left(\frac{\partial^2 w}{\partial x \partial y} \right)^2 - \left(\frac{\partial^2 w}{\partial x^2} \right) \left(\frac{\partial^2 w}{\partial y^2} \right) \right] \right\} dx\, dy$$

$$- \frac{1}{2} \int_{-a}^{a} \int_{-b}^{b} \left\{ F_x \left(\frac{\partial w}{\partial x} \right)^2 + 2 F_{xy} \frac{\partial w}{\partial y} \frac{\partial w}{\partial x} + F_y \left(\frac{\partial w}{\partial y} \right)^2 \right\} dx\, dy \qquad \dots(34)$$

Here, F_x, F_y and F_{xy}, are the in plane edge load and compressive load is considered as positive. For, finite element formulation the deflection in above expression needs to convert in terms of nodal displacements in the element.

Following the derivations in beam-column member (equations 18 to 22), the flexural and geometric stiffness for the plate element can be derived. Thus, the above expression can be derived to the following form using interpolation functions.

$$\pi = \frac{1}{2} \{d\}[k_F]\{d\} - \frac{F_x}{2} \{d\}^T [k_{Gx}]\{d\} - \frac{F_y}{2} \{d\}^T [kK_{Gy}]\{d\} - \frac{F_{xy}}{2} \{d\}^T [k_{Gxy}]\{d\} \quad \dots(35)$$

Here $[k_F]$ is the flexural stiffness matrix. The other stiffness matrices are analogous to the geometric stiffness matrices of the plate and can be expressed as:

$$k_{Gx} = \iint [N_x'] \{N_x'\}\, dx\, dy$$

$$k_{Gy} = \iint [N_y'] \{N_y'\}\, dx\, dy \qquad \dots(36)$$

$$k_{Gxy} = \iint [N'_x] \{N'_y\}\, dx\, dy$$

Where $[N'_x]$ and $[N'_y]$ indicate partial derivative of $[N]$ with respect to x and y respectively. Thus, the equation of buckling becomes:

$$[k_F]\{d\} - F_x [k_{Gx}]\{d\} - F_y [k_{Gy}]\{d\} - F_{xy} [k_{Gxy}]\{d\} = 0 \qquad \dots(37)$$

If the in plane loads have a constant ratio to each other at all time during their buildup, the above equation can be expressed as follows:

$$[k_F]\{d\} = P*(\alpha[k_{Gx}] + \beta[k_{Gy}] + \gamma[k_{Gxy}])\{d\} \qquad \dots(38)$$

The term is P* called the load factor and α, β and γ are constants relating the in-plane loads in the plate member. Solving the above expression, the buckling mode shapes are possible to determine.

4.9.1 Finite Elements in Fluid Mechanics and Ground Water Modeling

Governing Fluid Equations

Basically fluid is a material that conforms to the shape of its container. Thus, both the liquids and gases are considered as fluid. However, the physical behaviour of liquids and gases is very different.

The differences in behaviour lead to a variety of subfields in fluid mechanics. In case of constant density of liquid, the flow is generally referred to as incompressible flow.

The density of gases not constant and therefore, their flow is compressible flow. The Navier–Stokes equations are the fundamental basis of almost all fluid dynamics related problems. Any single-phase fluid flow can be defined by this expression. The general form of motion of a two dimensional viscous Newtonian fluid may be expressed as:

$$\frac{1}{\rho}p_i + \dot{v}_i + v_j\, v_{i,j} - \upsilon v_{i,ji} = f_i \qquad \text{...(1)}$$

Where,

υ- Kinematic viscosity.

ρ- Mass density of fluid.

v_i- Velocity components.

f_i - Body forces.

p - Fluid pressure.

The suffix i,j and i,ji are the derivatives along j and j & i direction respectively. The dot represents the derivative with respect to time. Neglecting non-linear convective terms, viscosity and body forces, and equation (1) can be simplified as:

$$p_i + \rho\dot{v}_i = 0 \qquad \text{...(2)}$$

Now, the continuity equation of the fluid is expressed by:

$$\dot{p} + \rho c^2\, v_{k,k} = 0 \qquad \text{...(3)}$$

Here, c is the acoustic wave speed in fluid. In the above expression, two sets of variables, the velocity and the pressure are used to describe the behaviour of fluid. Now it is possible to combine equation (2) and (3) to obtain a single variable formulation. For the small amplitude of fluid motion, one can assume:

$$v_i = \dot{u}_i \qquad \qquad ...(4)$$

Where, ui is the displacement component of fluid. To obtain single variable formulation, equation (4) may be substituted into equation (3) and one can get:

$$\dot{p} + \rho c^2 \dot{u}_{k,k} = 0 \qquad \qquad ...(5)$$

Integrating equation (5) with respect to time we have:

$$p = -\rho c^2 u_{k,k} \qquad \qquad ...(6)$$

Now differentiating the above expression with respect to xi following expression will be arrived:

$$p_i = -\rho c^2 u_{k,ki} \qquad \qquad ...(7)$$

Substituting the above in to equation (2) one can have:

$$\rho \dot{v}_i + \rho c^2 \dot{u}_{k,ki} = 0 \qquad \qquad ...(8)$$

Thus, equation (8) is expressed in terms of displacement variables only and known as displacement based equation.

Similarly, it is possible to obtain the fluid equation in terms of pressure variable only. Differentiating equation (3) with respect to time, the following expression can be obtained:

$$\ddot{p} + \rho c^2 \dot{v}_{k,k} = 0 \qquad \qquad ...(9)$$

Again, differentiating equation (2) with respect to xi, we have:

$$p_{,ii} + \rho \dot{v}_{i,i} = 0 \qquad \qquad ...(10)$$

From equations (9) and (10), the pressure based single variable expression can be arrived as given below:

$$\ddot{p} - c^2 p_{,i,i} = 0 \qquad \qquad ...(11)$$

The above expression is basically the Helmholtz wave equation for a compressible fluid having acoustic speed c:

$$\nabla^2 p - \frac{1}{c^2}\ddot{p} = 0 \qquad \qquad ...(12)$$

Thus, the general form of fluid equations of 2D linear steady state problems can be expressed by the Helmholtz equation. For incompressible fluid c becomes infinitely large. Hence for incompressible fluid, equation (12) can be written as:

$$\nabla^2 p = 0 \qquad \qquad ...(13)$$

For the ideal, irrotational fluid flow problems, the field variables are the streamline, ψ and potential φ functions which are governed by Laplace's equations:

$$\frac{\partial^2 \psi}{\partial x^2} + \frac{\partial^2 \psi}{\partial y^2} = 0$$

$$\frac{\partial^2 \phi}{\partial x^2} + \frac{\partial^2 \phi}{\partial y^2} = 0 \qquad \qquad ...(14)$$

Finite Element Formulation

Displacement and pressure based formulations will be derived using finite element method.

Displacement Based Finite Element Formulation

Consider the equation (8) which can be expressed only in terms of displacement variables:

$$\rho \ddot{u}_i - \rho c^2 u_{k,ki} = 0 \qquad \qquad ...(15)$$

Here, u is the displacement vector. Now, the weak form of the above equation will become:

$$\int_\Omega w_i \left(\rho \ddot{u}_i - \rho c^2 u_{k,ki} \right) d\Omega = 0 \qquad \qquad ...(16)$$

Performing integration by parts on the second terms, one can arrive at the following expression:

$$\int_\Omega w_i \rho \ddot{u}_i \, d\Omega - \int w_i \rho c^2 \, u_{k,k} \, d\Gamma + \int w_{i,i} \rho c^2 u_{k,k} \, d\Omega = 0$$

Or

$$\int_{\Omega} w_i \rho \ddot{u}_i \, d\Omega + \int_{\Omega} w_{i,i} \rho c^2 u_{k,ki} \, d\Omega = \int_{\Gamma} w_i \rho c^2 u_{k,ki} \, d\Gamma \qquad ...(17)$$

Now from earlier relation (equation 6) we have, $p = -\rho c^2 u_{k,k}$ .Thus, the above equation may be written as:

$$\int_{\Omega} w_i \rho \ddot{u}_i \, d\Omega + \int_{\Omega} w_{i,i} \rho c^2 u_{k,ki} \, d\Omega = -\int_{\Gamma} w_i p \, d\Gamma \qquad ...(18)$$

In case of fluid filled rigid tank, the weighting function wi must satisfy the condition wi ni = 0 on its rigid boundary. Therefore, the above equation will become:

$$\int_{\Omega} \left(w_i \rho \ddot{u}_i + w_{i,i} \rho c^2 u_{k,ki} \right) d\Omega = -\int_{\Gamma_p} w_i n_i p \, \Gamma \qquad ...(19)$$

For finite element implementation of the above expression, let consider the interpolation function as N and   as the nodal displacement vector. Thus:

$$u = N\bar{u} \text{ and } w = N\bar{w} \qquad ...(20)$$

Now the divergence of the displacement vector can be expressed as:

$$u_{i,i} = Lu = LN\bar{u} = B\bar{u} \qquad ...(21)$$

Where, $L = \begin{bmatrix} \dfrac{\partial}{\partial x} & \dfrac{\partial}{\partial y} \end{bmatrix}$ is the differential operator. Thus equation (19) may be written as:

$$w^T \int_{\Omega} \left[N^T \rho N \ddot{\bar{u}} + B^T \rho c^2 B\bar{u} \right] d\Omega = -w^T \int_{\Gamma_p} N^T n \bar{p} \, d\Gamma \qquad ...(22)$$

Or,

$$[M]\{\ddot{\bar{u}}\} + [K]\{\bar{u}\} = \{F\} \qquad ...(23)$$

Where,

$$[M] = \int_{\Omega} \rho [N]^T [N] d\Omega$$

$$[K] = \int_{\Omega} \rho c^2 [B]^T [B] d\Omega$$

$$[F] = -\int_{\Gamma_p} [N]^T n\{\overline{p}\} d\Gamma \qquad \text{...(24)}$$

Using equation (23), the displacements in fluid domain can be determined under external forces applying proper boundary conditions.

Pressure Based Finite Element Formulation

The Helmholtz equation (12) for a compressible fluid in two dimensions can be used to determine the pressure distribution in the fluid domain using finite element technique:

$$p, ii - \frac{1}{c^2}\ddot{p} = 0 \qquad \text{...(25)}$$

The weak form of the above expression can be written as:

$$\int_\Omega w_i \left(p, ii - \frac{1}{c^2}\ddot{p} \right) d\Omega = 0 \qquad \text{...(26)}$$

Now, performing integration by parts on the first term, the following expression can be obtained:

$$\int_\Gamma w_i p, i\, d\Gamma - \int_\Omega w_{i,i}\, p, i\, d\Omega - \int_\Omega \frac{1}{c^2} w_i\, \ddot{p}\, d\Omega = 0 \qquad \text{...(27)}$$

Thus,

$$\frac{1}{c^2}\int_\Omega w_i\, \ddot{p}\, d\Omega + \int_\Omega w_{i,i} p_{,i}\, d\Omega = \int_\Gamma w_i p_{,i} d\Gamma \qquad \text{...(28)}$$

Assuming interpolation function as N and $\overline{p}$ as the nodal pressure vector, the pressure (p) at any point can be written as: $p = N\overline{p}$ and similarly, $w = N\overline{w}$. The divergence of the pressure can be expressed as:

$$p, i = Lp = LN\overline{p} = B\overline{p}$$

where, $L = \left[\dfrac{\partial}{\partial x} \quad \dfrac{\partial}{\partial y}\right]$. Again:

$$w_{i,i} = Lw = LN\overline{w} = B\overline{w}$$

$$w_i\, \ddot{p} = \left[N\overline{w}\right]^T \left[N\ddot{\overline{p}}\right] = \overline{w}^T N^T N \ddot{\overline{p}} \qquad \text{...(29)}$$

Thus, equation (28) will become:

$$\frac{1}{c^2}\int_\Omega \overline{w}^T N^T N\ddot{\overline{p}}d\Omega + \int_\Omega \overline{w}^T B^T B\overline{p}d\Omega = \int_\Gamma w^T N^T \frac{\partial p}{\partial n}d\Gamma \qquad ...(30)$$

Or,

$$[E]\left\{\ddot{\overline{p}}\right\}+[G]\{\overline{p}\}=\{B\} \qquad ...(31)$$

Where,

$$[E]=\frac{1}{c^2}\int_\Omega [N]^T[N]d\Omega$$

$$[G]=\int_\Omega [B]^T[B]d\Omega$$

$$\{B\}=\int_\Gamma [N]^T \frac{\partial p}{\partial n}d\Gamma \qquad ...(32)$$

Applying boundary conditions, equation (31) can be solved to calculate the dynamic pressure developed in the fluid under applied accelerations on the domain.

Finite Element Formulation of Infinite Reservoir

Let consider an infinite reservoir adjacent to a dam like structure. In such case, if the dam is vibrated, the hydrodynamic pressure will be developed in the reservoir which can be calculated using above method.

For finite element analysis, it is necessary to truncate such infinite domain at a certain distance away from structure to have a manageable computational domain. The reservoir has four sides (Figure a) and as a result four types of boundary conditions need to be specified.

$$\{B\}=\{B\}_1 + \{B\}_2 + \{B\}_3 + \{B\}_4 \qquad ...(33)$$

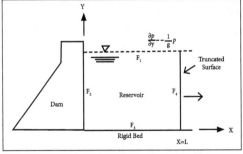

(a) Reservoir and its boundary conditions.

At the Free Surface (Γ_f)

Neglecting the effects of surface waves of the water, the boundary condition of the free surface may be expressed as:

$$p(x, H, t) = 0 \qquad \qquad ...(34)$$

Here, H is the depth of the reservoir. However, sometimes, the effect of surface waves of the water needs to be considered at the free surface. This can be approximated by assuming the actual surface to be at an elevation relative to the mean surface and the following linearized free surface condition may be adopted:

$$\frac{1}{g}\ddot{p} + \frac{\partial p}{\partial y} = 0 \qquad \qquad ...(35)$$

Thus, the above expression may be written in finite element form as:

$$\{B_1\} = \frac{\partial p}{\partial n} = -\frac{1}{g}[R_1]\{\ddot{p}\} \qquad \qquad ...(36)$$

In which,

$$[R_1] = \int_{\Gamma_1}[N]^T[N]d\Gamma \qquad \qquad ...(37)$$

At the Dam-reservoir Interface (Γ_s)

At the dam-reservoir interface, the pressure should satisfy:

$$\frac{\partial p}{\partial n}(0, y, t) = \rho a e^{i\omega t} \qquad \qquad ...(38)$$

Where, $ae^{i\omega t}$ is the horizontal component of the ground acceleration in which, ω is the circular frequency of vibration and $i = \sqrt{-1}$, n is the outwardly directed normal to the elemental surface along the interface.

In case of vertical dam-reservoir interface $\partial p/\partial n$ may be written as $\partial p/\partial x$ as both will represent normal to the element surface. For an inclined dam-reservoir interface $\partial p/\partial x$ cannot represent the normal to the element surface.

Therefore, to generalize the expressions $\partial p/\partial n$ is used in equation (38). If $\{a\}$ is the vector of nodal accelerations of generalized coordinates, $\{B2\}$ may be expressed as:

$$\{B_2\} = -\rho\,[R_2]\{a\} \qquad \qquad ...(39)$$

Where, $[R_2] = \Sigma \int\limits_{\Gamma_2} [N_r]^T [T][N_d] d\Gamma$ ...(40)

Here, [T] is the transformation matrix for generalized accelerations of a point on the dam reservoir interface and [Nd] is the matrix of shape functions of the dam used to interpolate the generalized acceleration at any point on their interface in terms of generalized nodal accelerations of an element.

At the Reservoir Bed Interface (Γ_b)

At the interface between the reservoir and the elastic foundation below the reservoir, the accelerations should not be specified as rigid foundation because they depend on the interaction between the reservoir and the foundation. However, for the sake of simplicity, the reservoir bed can be assumed as rigid and following boundary condition may be adopted.

$$\{B_3\} = \frac{\partial p}{\partial \eta} \ (x,0,t) = 0 \qquad ...(41)$$

At The Truncation Boundary (Γ_t)

The specification of the far boundary condition is one of the most important features in the finite element analysis of a semi-infinite or infinite reservoir. This is due to the fact that the developed hydrodynamic pressure, which affects the response of the structure, is dependent on the truncation boundary condition.

The infinite fluid domain may truncate at a finite distance away from the structure for finite element analysis satisfying Summerfield radiation boundary condition. Application of Summerfield radiation condition at the truncation boundary leads to:

$$\{B_4\} = \frac{\partial p}{\partial x} \ (L,y,t) = 0 \qquad ...(42)$$

Here, L represents the distance between the structure and the truncation boundary. Thus, the hydrodynamic pressure developed on the dam-reservoir interface can be calculated under external excitation by the use of finite element technique.

Finite Elements in ground water modeling

Models

A model is a tool designed to represent a simplified version of reality. Given this broad definition of a model, it is evident that we all use models in our everyday lives.

For example, a road map is a way of representing a complex array of roads in a symbolic form so that it is possible to test various routes on the map rather than by trial and error while driving a car. Similarly, groundwater models are also representations of reality and if properly constructed, can be valuable predictive tools for management of groundwater resources.

For example, using a groundwater model, it is possible to test various management schemes and to predict the effects of certain actions. Of course, the validity of the predictions will depend on how well the model approximates field conditions.

Good field data are essential when using a model for predictive purposes. However, an attempt to model a system with inadequate field data can also be instructive as it may serve to identify those areas where detailed field data are critical to the success of the model. In this way, a model can help guide data collection activities.

Types of Groundwater Models

Several types of models have been used to study groundwater flow systems. They can be divided into three broad categories. Sand tank models, analog models, including viscous fluid models and electrical models and mathematical models, including analytical and numerical models.

A sand tank model consists of a tank filled with an unconsolidated porous medium through which water is induced to flow. A major drawback of sand tank models is the problem of scaling down a field situation to the dimensions of a laboratory model.

Phenomena measured at the scale of a sand tank model are often different from conditions observed in the field and conclusions drawn from such models may need to be qualified when translated to a field situation.

The flow of groundwater can be described by differential equations derived from basic principles of physics. Other processes, such as the flow of electrical current through a resistive medium or the flow of heat through a solid, also operate under similar physical principles.

In other words, these systems are analogous to the groundwater system. The two types of analogs used most frequently in groundwater modeling are viscous fluid analog models and electrical analog models.

Viscous fluid models are known as Hele-Shaw or parallel plate models because a fluid more viscous than water (for example, oil) is made to flow between two closely spaced parallel plates, which may be oriented either vertically or horizontally.

Electrical analog models were widely used in the 1950s before high-speed digital computers became available. These models consist of boards wired with electrical networks of resistors and capacitors.

They work according to the principle that the flow of groundwater is analogous to the flow of electricity. This analogy is expressed in the mathematical similarity between Darcy's law for groundwater flow and Ohm's law for the flow of electricity.

Changes in voltage in an electrical analog model are analogous to changes in ground-water head. A drawback of an electrical analog model is that each one is designed for a unique aquifer system. When a different aquifer is to be studied, an entirely new electrical analog model must be built.

A mathematical model consists of a set of differential equations that are known to govern the flow of groundwater. Mathematical models of ground-water flow have been in use since the late 1800s.

The reliability of predictions using a groundwater model depends on how well the model approximates the field situation. Simplifying assumptions must always be made in order to construct a model because the field situations are too complicated to be simulated exactly.

Usually the assumptions necessary to solve a mathematical model analytically are fairly restrictive. For example, many analytical solutions require that the medium be homogeneous and isotropic.

To deal with more realistic situations, it is usually necessary to solve the mathematical model approximately using numerical techniques. The subject here is the use of numerical methods to solve mathematical models that simulate groundwater flow and contaminant transport. We consider two types of models, finite difference models and finite element models. In either case, a system of nodal points is superimposed over the problem domain.

For example, consider the problem shown in the figure (1). The problem domain consists of an aquifer bounded on one side by a river (Figure (1a)). The aquifer is recharged really by precipitation, but there is no horizontal flow out of or into the aquifer except along the river.

Two finite difference representations of the problem domain are illustrated in the figures the 1(b) and 1(c) and a finite element representation is shown in the figure (d).

The concept of elements is fundamental to the development of equations in the finite element method.

Triangular elements are used in the figure (a), but quadrilateral or other elements are also possible. In the finite difference method, nodes may be located inside cells (Figure 1b) or at the intersection of grid lines (Figure 1c).

The finite difference grid shown in the figure (1b) is said to use block-centered nodes, whereas the grid in the figure (1c) is said to use mesh-centered nodes. Aquifer properties and head are assumed to be constant within each cell in the figure (1b).

In the figure (1) nodes are located at the intersections of grid lines and the area of influence of each node is defined following one of several different conventions.

Regardless of the representation, an equation is written in terms of each nodal point because the area surrounding a node is not directly involved in the development of finite difference equations.

The goal of modeling is to predict the value of the unknown variable (for example, groundwater head or concentration of a contaminant) at nodal points. Models are often used to predict the effects of pumping on groundwater levels.

For example, consider the aquifer shown in the figure (1). In this example, a model could be used to predict the effects of pumping the three wells in the well field on water levels in the four observation wells or to predict the effects of installing additional pumping wells.

The model could also be used to determine how much water would be diverted from the river as a result of pumping. However, before a predictive simulation can be made, the model should be calibrated and verified.

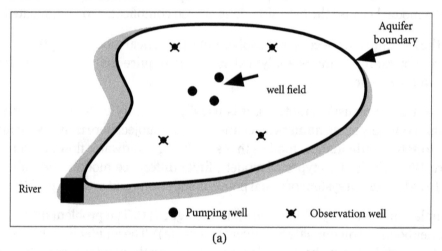

(a)

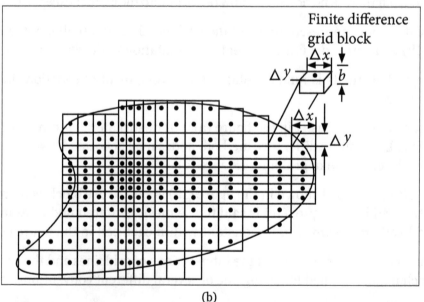

(b)

(1) Finite difference and finite element representations of an aquifer region. (a) Map view of aquifer showing well field, observation wells and boundaries. (b) Finite difference grid with block-centered nodes, where Δx is the spacing in the x direction, Δy is the spacing in the y direction and b is the aquifer thickness.

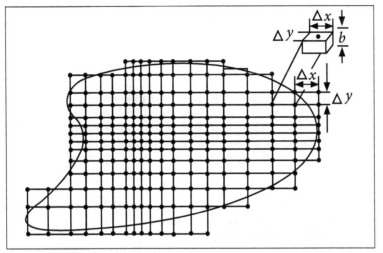

(c) Finite difference grid with mesh-centered nodes.

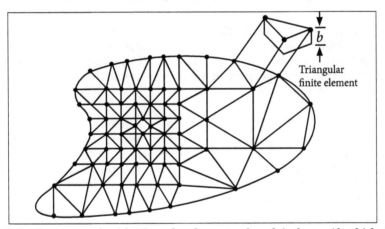

(d) Finite element mesh with triangular elements where b is the aquifer thickness.

Permissions

Index

L
Linear Equations, 35, 38
Linear Polynomial, 91
Linear System, 12
Linearity, 2
Longitudinal Strain, 142

M
Magnetic Field, 70
Material Behavior, 76-77
Mathematics, 8, 151
Matrix, 5-7, 25, 31-33, 38, 43, 45-46, 48, 68-69, 76-79, 81, 83-84, 167-169, 172, 180-182, 186-187, 191-192, 196-197, 200, 202-203, 205, 207-209, 218-219, 224-225, 227, 229, 237
Matrix Equation, 31, 99, 203, 205
Matrix Form, 48, 78-79, 83, 101, 116, 126, 144, 147, 154, 180, 203
Means, 3, 20, 64-67, 85, 152, 210
Meshes, 86

O
Ode, 34

P
Partial Differential Equations, 1-2, 8, 18, 85
Poisson, 35
Pollution, 71-72
Polynomial, 4, 68-69, 90-91, 93-97, 100-101, 103, 112-113, 118-119, 135, 137, 151-153, 156, 227
Power Series, 97
Precipitation, 239

Q
Quadratic Polynomial, 95-96, 113

R
Red Blood Cells, 214
Reliability, 239

Reservoir, 235, 237
Resistance, 81, 209-210
Rigid Body, 54, 79, 90-91, 161, 215
Rotation, 12, 90, 94-95, 186, 191, 194, 196, 200, 207, 210, 215, 218

S
Square Matrix, 83, 209
Stability Analysis, 65, 71, 224, 229
Structural Analysis, 1, 64-66, 76-77, 80-81, 84
Subscripts, 52, 55
Subspace, 9-10
Suffix, 230

T
Taylor Series, 20, 32
Theorem, 8, 26, 77, 80, 83
Thermal Analysis, 71
Time Derivative, 23
Translation, 94-95
Triangle, 10, 58, 75, 95, 99, 103, 109, 119, 124, 132, 139-140, 144, 163, 168

V
Variables, 7, 18, 21, 34, 36, 68-69, 77, 81, 85, 91, 93-95, 231-232
Velocity, 35-36, 68, 90, 230-231
Viscosity, 230
Visualization, 2
Voltage, 72, 239

W
Wave Equation, 232

Printed in the USA
CPSIA information can be obtained
at www.ICGtesting.com
JSHW052127021123
51365JS00005B/22